ALIEN THREAT FROM THE MOON

Dylan Clearfield

ISBN 978-0-930472-30-6

CONTENTS

Introduction

Our Moon is currently occupied by aliens and they are hostile toward any other visitors. That is the primary reason why no new missions as of this writing have been sent to the lunar surface since the last astronauts left so many decades ago. Photographs reveal the locations of alien activities and anomalies which are in reality quite frightening. What you are going to see in the following pages will scare you. It should. The photographs reveal not only an alien presence but what that presence suggests about humankind and its greatly inferior location on the scale of advancement.

All of the photographs were taken by NASA. None of them have been altered in anyway, except in those places where NASA itself was attempting to either conceal or create evidence. In those cases, the fact of the attempted cover-ups spoke more loudly than would have the actual evidence if we'd been allowed to see it. It is only because of the sheer immensity of the number of photographs that NASA has taken that those few which you are about to see escaped purging. Even NASA could not parse through every last photo in the attempt to erase all evidence of alien activity.

The photographs about to be revealed show clear signs of alien use of the Moon as well as the remains of the original or "native" lunar inhabitants. You are not going to be asked to stress your powers of vision to the ultimate limit to visualize a fuzzy object as being a wall or bridge or pipe. NASA has provided very sharp photographs of the Moon. Close ups are used to observe many of the objects. Fortunately, one of the most powerful scanners in existence was available to give clarity to these objects. The identity of some of the objects is obvious, some require speculation and some – while clearly visible – defy explanation. They are there, but they simply escape definition.

Most alien sites exist on the "Near side" probably for logistical reasons, allowing them to be in constant contact with earth. If they had wanted to hide from us they could easily have done so on the "Dark side" but that probably would have caused them more communications problems than would have been worth the effort.

There also are areas of former "native" habitation on the Moon as well as evidence of the remains of flora and fauna. People used to live here? There used to be plants and animals on the Moon, too? Yes. Even the great astronomer Sir John Herschel – among several others – noted this himself. It seems clear that some type of flora and fauna once roamed the lunar surface ages ago. Evidence of this will be shown at the proper time.

There clearly are the remains of small villages or even cities on the Moon which had belong to the original inhabitants, not the aliens. As a trained archaeologist, if I would observe the same features on earth as exist on the Moon regarding this subject I would without question categorize them as cities seen from above. Photographs can and will prove their existence, but they cannot date their time period.

A word about the type of photographs to be used must be mentioned. Almost all of the photographs of the lunar surface are of the highest quality that were taken by Apollo astronauts either from close orbit or while on the surface. They are incredibly sharp and not like the usual ones to be found on the internet. In addition, one of the most powerful scanners and magnifiers that exists was used to develop close ups to within a distance of three hundred feet. Great detail can be obtained from this close of a distance. These are the sharpest photographs that exist at the time of the writing of this book.

In addition, many of the photos have been taken from this author's vast historical library of lunar images (dating back to the lunar atlas of 1903). What this means is that there is access to the original NASA photos taken of the moon decades ago, photos which have escaped the later alterations made to them by the Space Agency.

The evidence of alien presence on the moon can be found in many of these earlier pictures and some will be compared to the more modern NASA versions which have been censored. In this book the reader will be allowed to see what the real Moon looks like and who is now and who was in the past inhabiting it.

Many of the communications between NASA and its astronauts were kept secret by using special transmitters and receivers. For decades this has been considered a mere conspiracy theory. In the chapter titled "Encounters between astronauts and aliens" proof is supplied that the coded communications between the astronauts on the surface of the Moon and NASA really did happen and were faithfully recorded by amateur radio operators on earth.

While these secret discussions were taking place on the surface, photos were being taken from orbit, revealing the sights that the astronauts were describing to NASA. These images will be shown here along with the actual recordings that accompanied them! The photographs exist as part of the author's historical lunar library and have been removed from public view by the Space Agency. But these images reveal that what the astronauts were reporting to NASA in coded conversations was hidden from the public.

Mysterious happenings in Mare Imbrium

After many years of taking photographic records of Mare Imbrium it appears that changes have regularly taken place on portions of its landscape. Changes of an unnatural and startling nature. And what they imply is that someone is using the lunar terrain as a type of message board as well as a major base of operations. Right in front of everyone's eyes!

If such changes have actually occurred why haven't they been widely reported? After all, Mare Imbrium is a major feature on the Moon. Maybe these occurrences haven't been noticed precisely because of this area's high visibility. It is often the evidence that is the most obvious that is the most overlooked. Below is a photo of the Mare Imbrium Region.

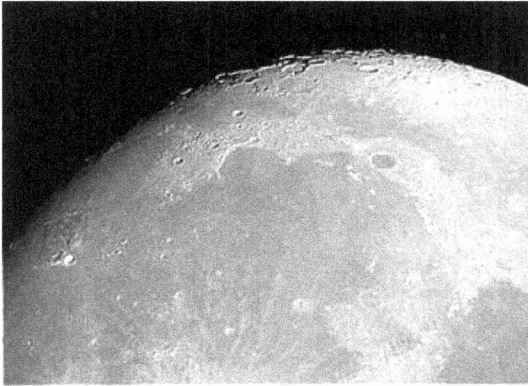

One place where unusual activity has been occurring is within a specific "natural" landform feature in the area near the upper edges of the mare. Below is a closer view of the area. Of concern is that group of rocky formations at the top center and left of the photo.

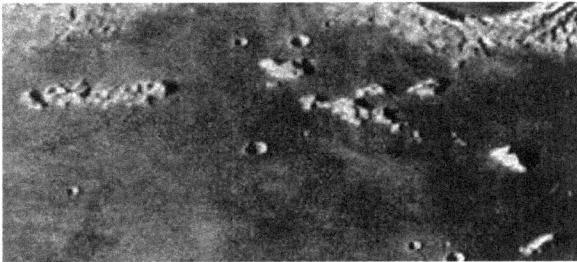

Note in particular the grouping on the left which is highlighted below.

Next focus more specifically on the end piece on the right.

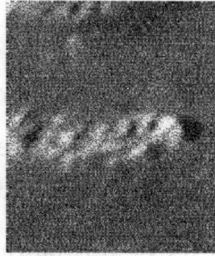

Why is this important? Primarily because of the unusual shape of this end piece of rugged stone. Is it not shaped like a broccoli crown? Or, more threateningly, a mushroom cloud?

Next is a nearby location where a similar mushroom cloud symbol is located. It is in another rock formation which is to the

right of the group with which this report began – also in Mare
Imbrium.

A simple coincidence or a coded signal to someone who for
some reason must be shown these types of messages in the form of
colossal rocky formations?

Couldn't this just be a coincidence? A geological freak?
Probably not. And that is because of the recurrence of this same
type of image in other locations related to the Moon and UFO
activity. For example, in more than one crater on the Moon there
exists an outcrop of rocks that looks very similar to the one
pictured next.

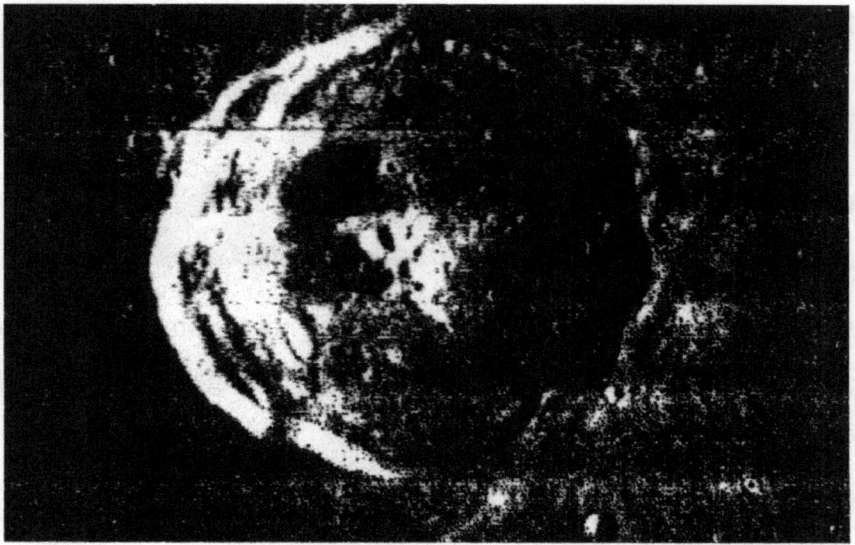

Below is a closer view.

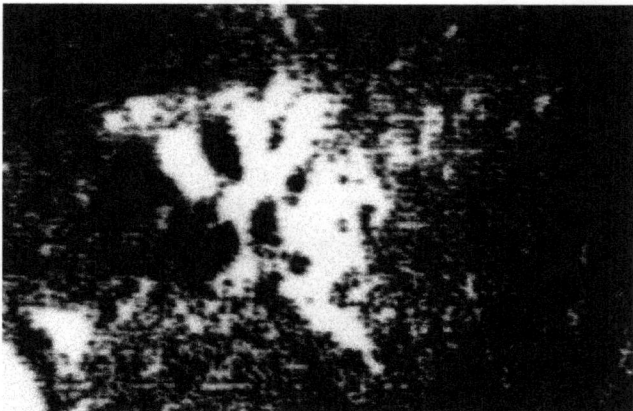

Whether or not this symbol represents an image which is coded to look like a broccoli crown or is actually meant to be a mushroom cloud, its appearance at many places across the lunar surface is at least mystifying. But there is more. This same symbol has appeared in another shocking location – broadcast to the world on the side of a UFO!

Next is the picture of the mushroom cloud image that was flashed on the side of a UFO before an amazed crowd of onlookers in Brazil in 1990.

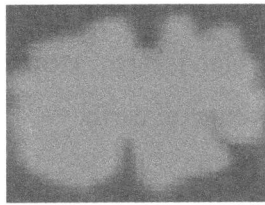

What follows is a side-by-side comparison of the symbol flashed on the UFO and one of the formations in one of the rocky outcrops in Mare Imbrium

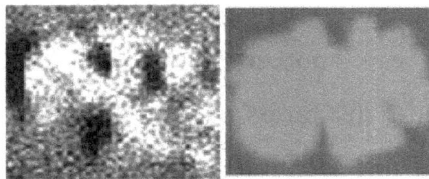

Thus far, only one of the oddly shaped rocky outcrops in Mare Imbrium has been considered. But the rocky outcrop on the

right side when viewed in its entirety also raises questions. In the photograph that follows, this rocky formation is shown from different angles. Does this not look like some type of writing?

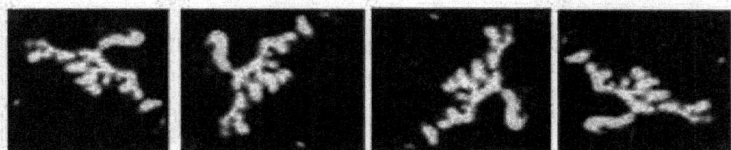

This could in fact represent four different messages depending on which direction from which they are viewed from above. And they look even less like natural formations when seen this way. The point being made is that since there are clearly coded messages on the lunar landscape, extraterrestrials seem to be involved in upcoming events on earth.

Someone seems to be changing the terrain on the surface of Mare Imbrium, using the landforms as symbols. One factor that appears to be likely is that whoever is altering the landforms in Mare Imbrium must be highly technologically advanced in order to be able to accomplish this.

But there is another nearby location on the Moon where just the opposite, insofar as technological advancement, seems to be the case. In this location it appears that a landform had been created some time in the distant past possibly in order to draw the attention of airborne aliens or else as a way to contact god-like beings.

Had there at one time been an ancient native race of people populating the Moon who, like our human ancestors, may have attempted to contact divine beings in the sky through formations made of geological material?

Or maybe the landform about to be examined performed a completely different function. Maybe it was to mark the location of a significant historical event or of some other place of great importance like the following plaque on earth.

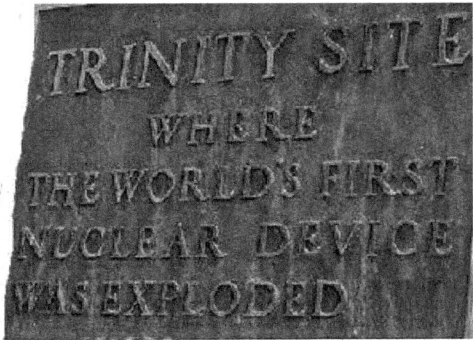

Or, to highlight another place:

Or perhaps one of more historical significance:

It clearly isn't unusual for a society to honor a special site or location with a prominent sculpture. The lunar landform that follows is located near the Sinus Iridium area of the Moon on a plain about 50 km east of the Crater Maupertuis. Sinus Iridium is a close neighbor of Mare Imbrium, being an extension of it to the northwest but not very popular among astronomers and NASA.

In the picture below, the mysterious writing near the Crater Maupertuis is located in the yellow oval to the north of Sinus Iridium.

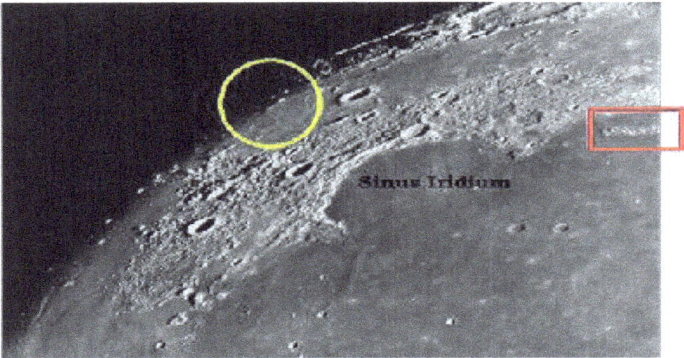

What follows is the writing on the lunar surface that appears about two hundred miles to the northwest of the figures found in Mare Imbrium. It appears to be some form of writing, but in an unknown script. It almost looks like the word Greene.

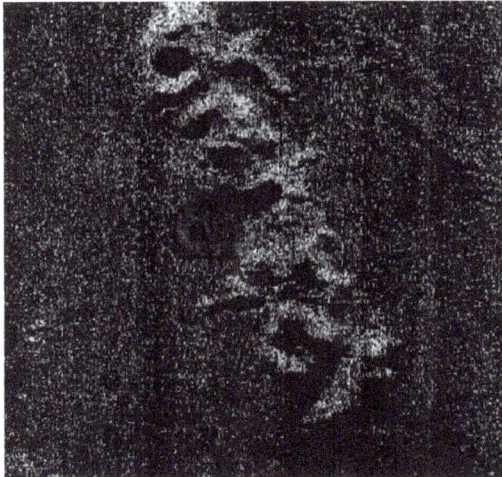

It is enhanced below for better readability.

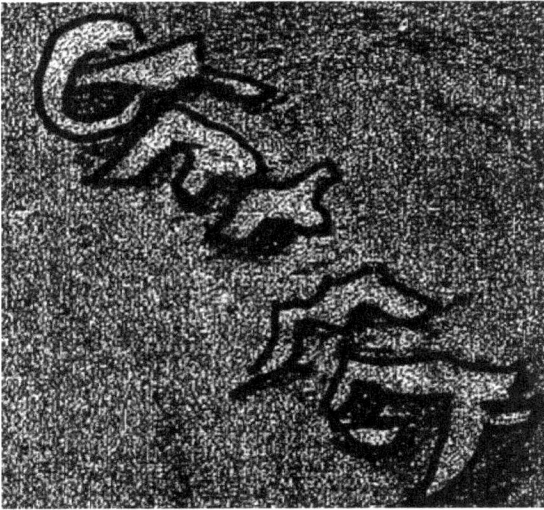

This image is basically non-existent among NASA's photographic atlases of the Moon. It was discovered by this author in a little known publication about lunar geology that was published decades ago by an association affiliated with NASA. It is unlikely that NASA intended for this image to have been widely published if at all.

How would it be possible for a design like this to have formed on a lunar surface which has no active processes of erosion? Note the first "letter" of the formation which looks like a C or G. How could that have naturally formed? Beside it is what looks like a small case r. How likely would it be for two such rocky outcrops to naturally attain the form of letters?

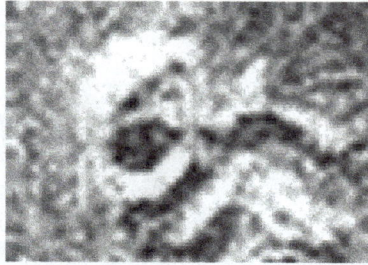

Also notice the form that is at the top right of the C (or G). Does this not look like a marlin fish attempting to jump through a tiny hoop?

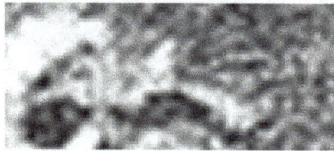

But, considering the message in its totality, might the symbols represent a type of distress signal, a plea for help like an alien form of SOS? Or could the letters spell out a large commercial sign like SEARS?

Or maybe even a NO TRESPASSING sign.

Hotspots of Alien Activity on the Moon

The most concentrated area of alien activity is in the Mare Imbrium area of the Moon. This happens to be one of the most visible and easily observed areas on the Moon. From our perspective on earth, part of the Mare Imbrium region would occupy the area to the left which is encircled in red.

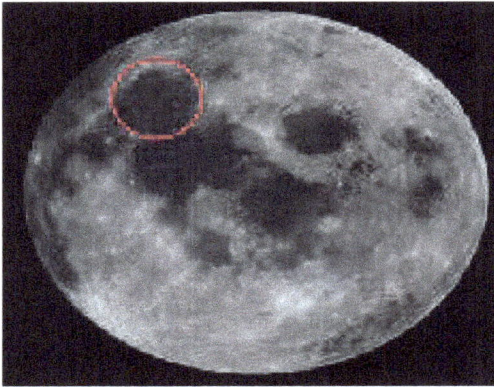

Below is a wide angle view of the entire Mare Imbrium area.

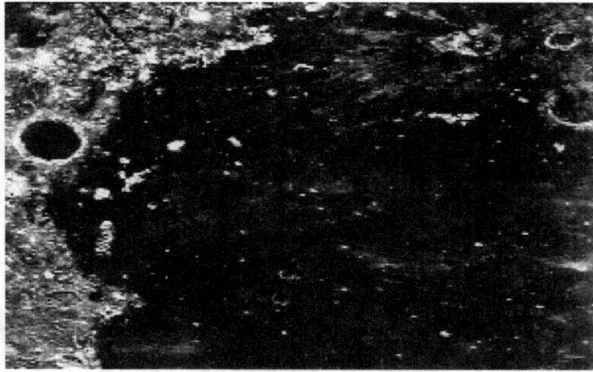

This seems a logical location for the majority of alien activities to take place. The ground is relatively flat and not overwhelmed by craters and the mountains are in the distance allowing easy access for flying craft.

Even though this location is relatively devoid of craters and obviously very smooth not a single Apollo Mission landed in Mare Imbrium. WHY NOT? Could it be because the aliens warned us not to land there? Otherwise, it makes no sense that not a single mission would have been sent there. Apollo 15 got closest to Mare Imbrium but did not venture inward.

Mare Imbrium again. The coded messages are to the extreme left. The area in the red circle is another particularly important location.

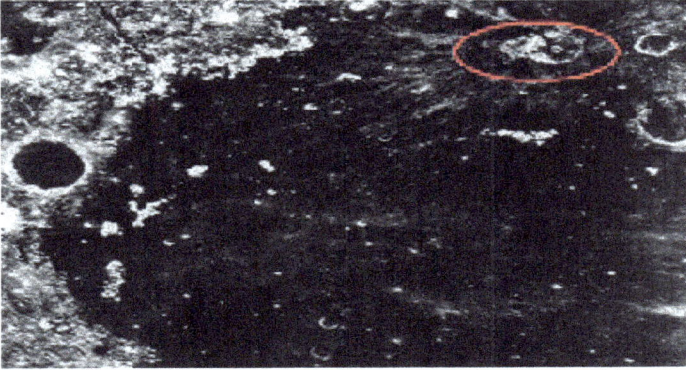

This appears to be a significant alien technological center. It is very likely that this is where aircraft, spacecraft and lunar traversing machines are built and fixed. A closer view follows.

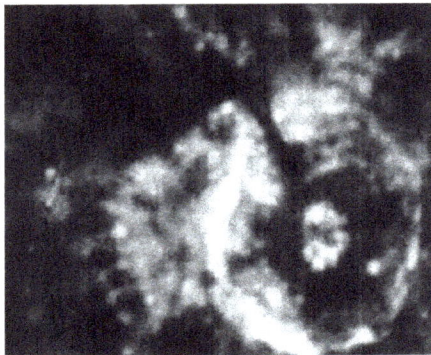

Note the areas enclosed in red below. These are similar to "gates" that might be found at an airport on earth. In this case, however, these are work areas which are supplied with power and all of the other necessities from a central core location to the right.

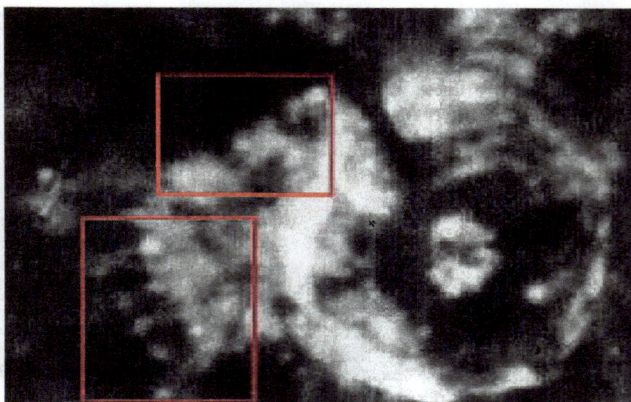

In the very center of this complex is an extremely high, tower like structure. It is probably from where all of the activities are controlled. Admittedly, its upward extension is difficult to see.

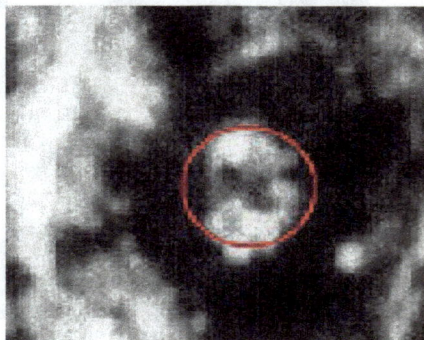

Before you begin to consider that this is all mere speculation, first take a look at a photograph of one of the most common types of UFO that are reported on earth. The cylinder shaped UFO. These are genuine photographs taken in the skies of earth.

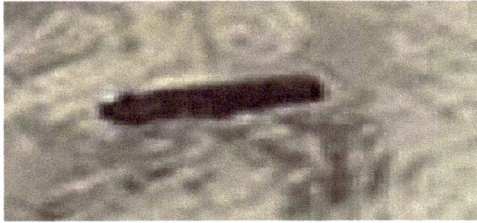

AND

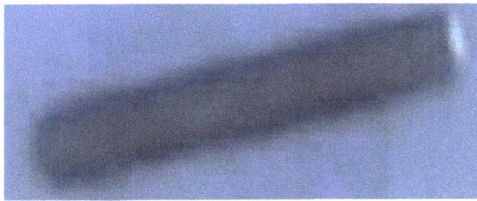

Two of the best photos of genuine tube-shaped UFOs. On the lunar surface, the same type of UFO is found on the ground at the edge of the alien technological center.

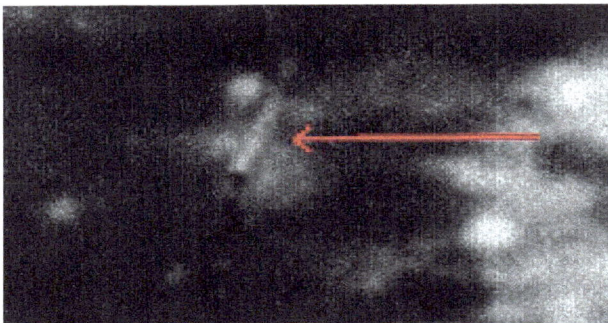

Notice also that there seems to be some type of black fluid pouring out of the cigar shaped object. Later, this same type of craft will be seen on another part of the Moon under very unusual circumstances. Below is a close up of the one that is currently being considered.

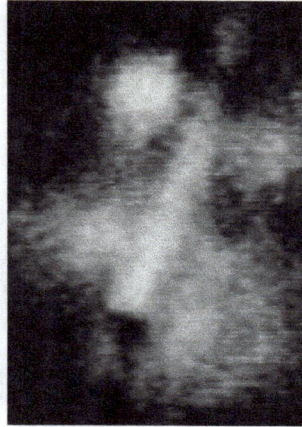

This is just one of a number of alien craft discovered in Mare Imbrium. Another mysterious grouping is pointed out in the next photo.

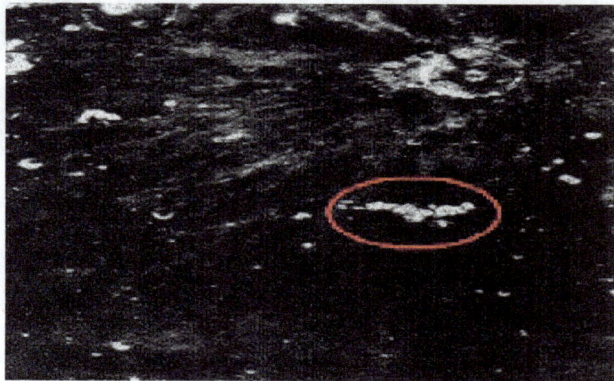

This area is to the south of the one just investigated. There are at least two alien spacecraft in this group within the above red circle. A close up of this location follows. The photos are of the same location but from different vantage points. Extreme magnification was necessary which accounts for the blurriness.

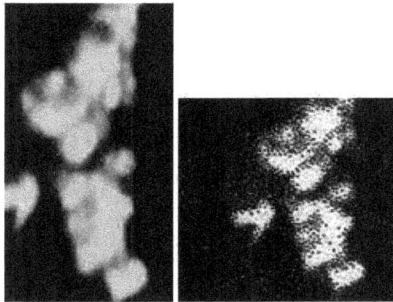

Enclosed within the red squares below are the two alien aircraft. One of them is of the typical saucer-shaped variety:

And the second one is a unique winged variety which has three turbo jet extensions in the back.

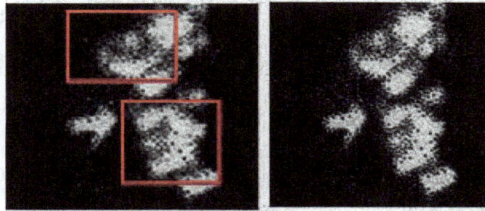

It is the belief that these two objects are in this location to receive some form of repair or simply to be refueled.

The next photographs are absolutely critical to the study of alien activity on the Moon. And they cannot be ignored. The photos you are about to see were taken by NASA almost exactly one year apart. Do you notice the difference? On the right is one year later! This is in the same repair zone just observed above.

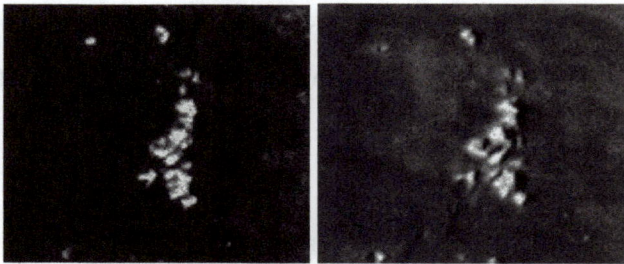

In the photo on the right, the saucer shaped object has been opened on the right side as if some form of overhaul is being performed on it. The winged aircraft that was in the photo on the left is now gone! This is the same location on the Moon one year apart. See the closer view next:

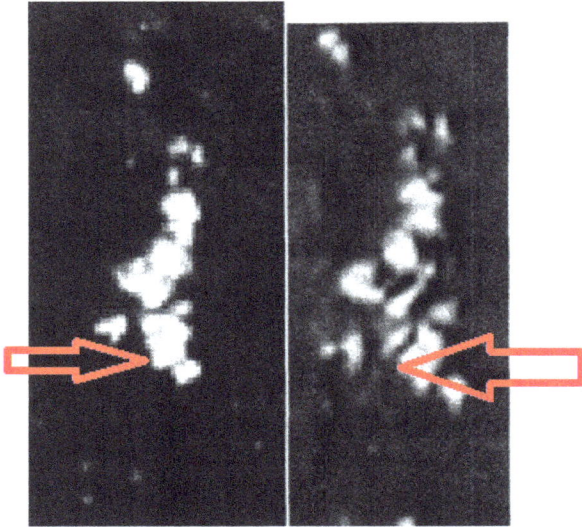

Obviously the images are extremely difficult to see but that doesn't prevent a viewer from noticing the differences. The picture below was provided with arrows to help locate the suspect areas.

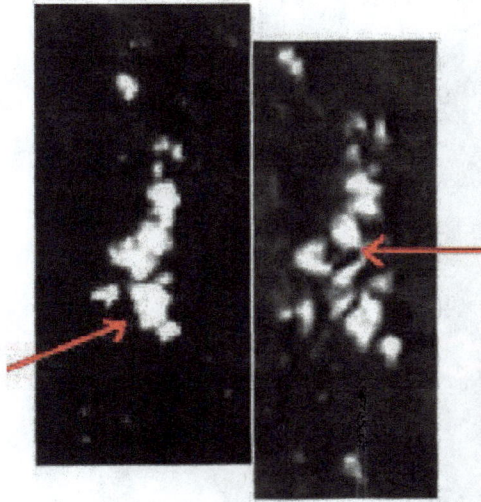

What became of the craft on the left? In the photo on the right it is gone and in its place are what seem to be two gigantic broken boulders of a different shape. Were these boulders supporting the craft previously? Had the vehicle actually crashed there and simply left the area after being repaired?

And what about the saucer? (Red arrow on the right). It is still there but not in the same condition as before. It looks like it had been cut open with the portion that was removed being laid on its back next to it. Note the darkened area. Both photos – a year apart – were taken at the same time of day.

No matter what type of explanation is given for these photos it is an absolute fact that they show that this location on the

Moon has changed. It could not have happened by means of weathering.

There is one other area in Mare Imbrium that deserves examination. It is located here:

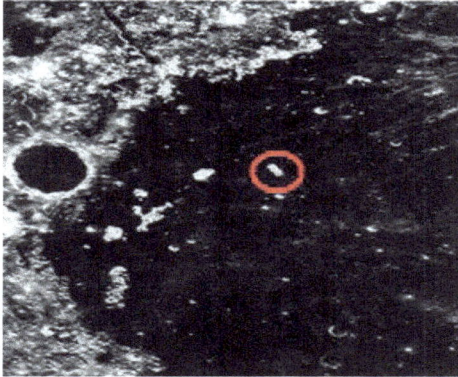

A closer look will help. Below is the enlarged view of this strange object circled in red.

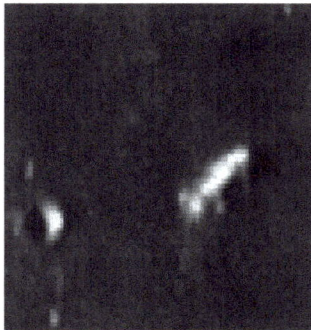

Whatever this object is, it is clearly being supported on legs of some type. It appears to be some form of upraised lunar device whose purpose can only be guessed at. Maybe saucers are placed beneath it and it works on them from above. Maybe this is a device that supplies various types of fluids for machines here that work on the surface. The point is that whatever that object is, it is obviously raised above the surface and is supported on legs.

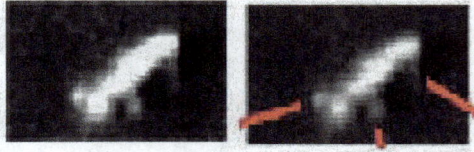

For a supposedly dead world there is a great deal of activity taking place on the Moon. And Mare Imbrium is not the only place. The object just shown – or most likely its duplicate – will be seen at another location on the Moon later in this study.

Next to be examined is Crater Copernicus. The aliens are quite busy here. NASA doesn't want this known. However, NASA has a big problem. Copernicus is a huge crater and NASA cannot simply deny its existence.

One way to accomplish hiding a crater in plain sight is by showing that crater from different angles and lighting. This will confuse people as to what the crater really looks like. The prime example is Crater Copernicus. First, the view that NASA wants you to see.

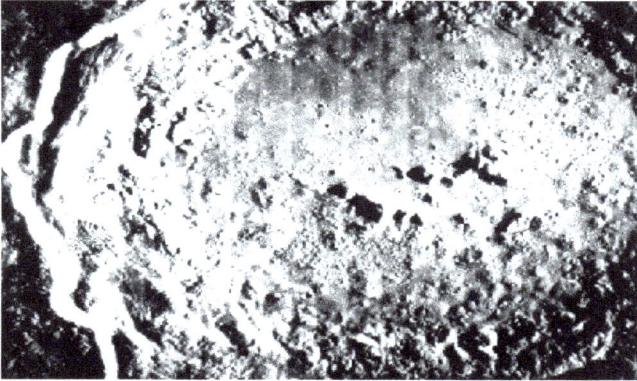

In this view, the crater is mostly bleached out by the sunlight. In truth, there is almost as much alien activity taking place in Copernicus as there is in Mare Imbrium. Even in the above bleached out version some of the alien influence is still visible. And it will be found throughout the crater in other, clearer pictures.

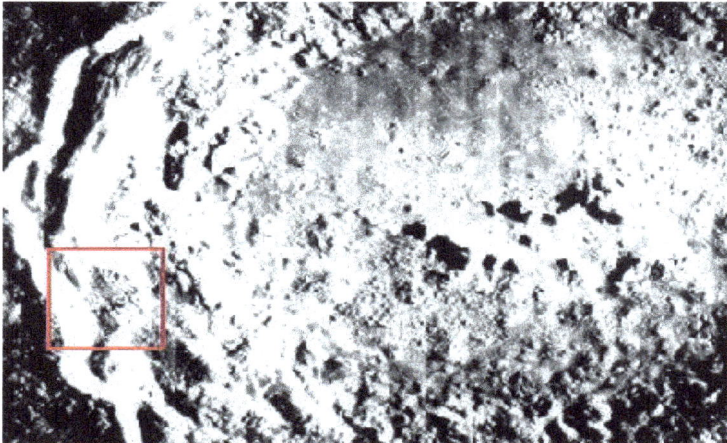

That area in red is not simple crater material. Take a closer look at it.

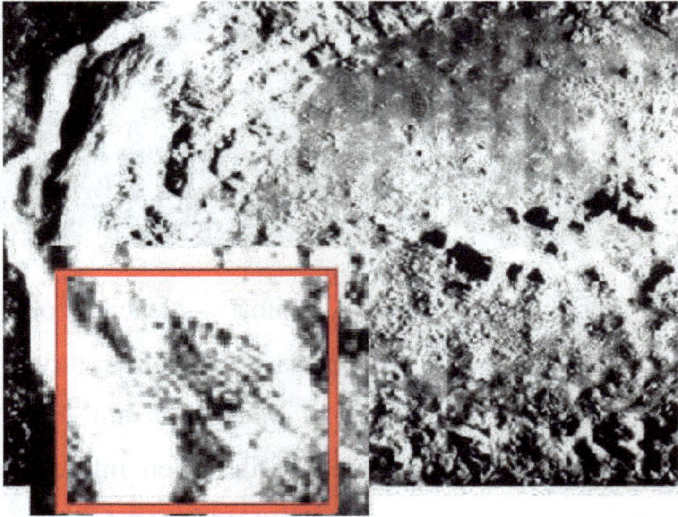

Even closer:

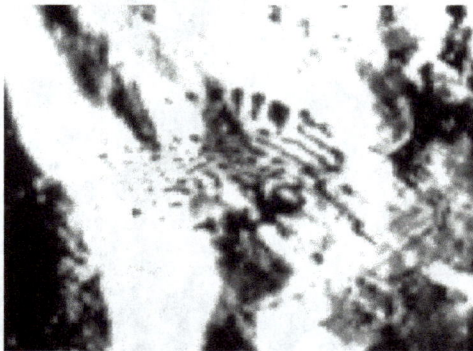

This appears to be some form of metallic mesh woven into the top of the crater rim. Other than this, its purpose or definite identification is as yet unknown.

When a clearer view of the Copernicus crater is shown it becomes more difficult to disprove alien activity than to prove it. Here is a more realistic photograph of the crater rather than the bleached out one.

First to be studied will be more of the rim on the left side of the crater. A close up reveals bizarre honeycomb like features that clearly aren't natural.

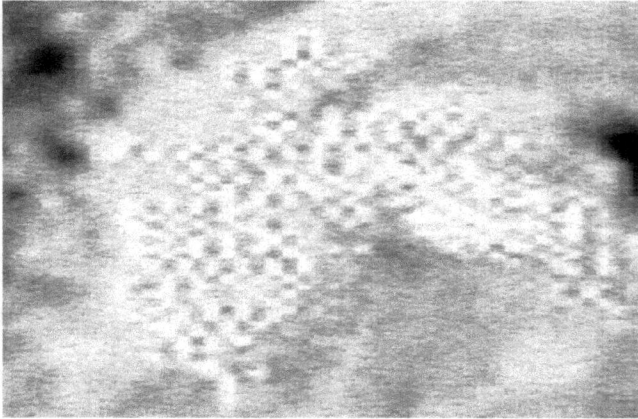

There's more of the same in the bottom of the crater. This is atop an elevated section in the middle of the crater.

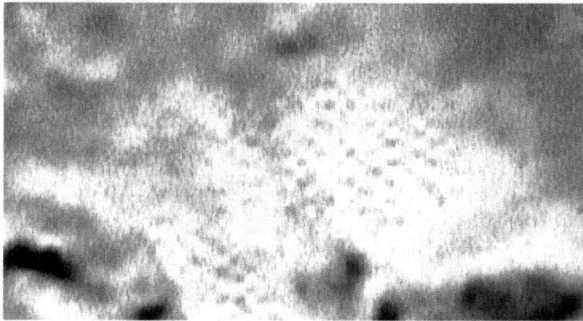

This material is different from the mesh on the rim. It looks very much like perforated metal, a type of construction material that is widely used on earth.

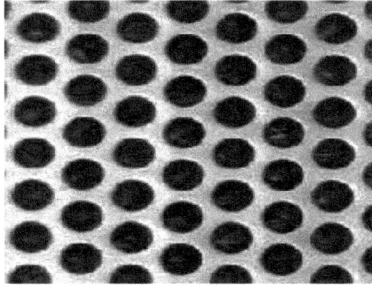

Perforated metal would be an ideal way of lining the sides of a crater to strengthen it for whatever uses it was intended. It's a common product that even aliens would use. But there's a lot more taking place in Crater Copernicus. And one of the most important areas of activity is in the exact center of the crater. It couldn't be seen in the bleached out photograph but the object in the middle of Copernicus can be definitely seen in the better quality picture. This is a close up of the center of the crater.

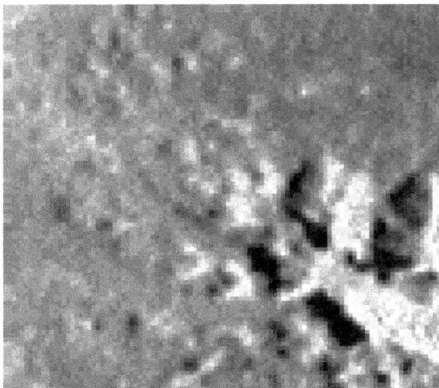

Near the center of the picture is an odd looking artificial type object. It is circled in red below and is shaped like a silver cartridge like those used in revolvers.

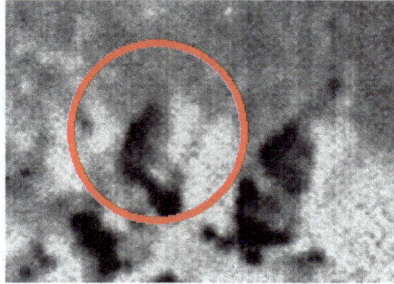

It's actually easier to see from the distance:

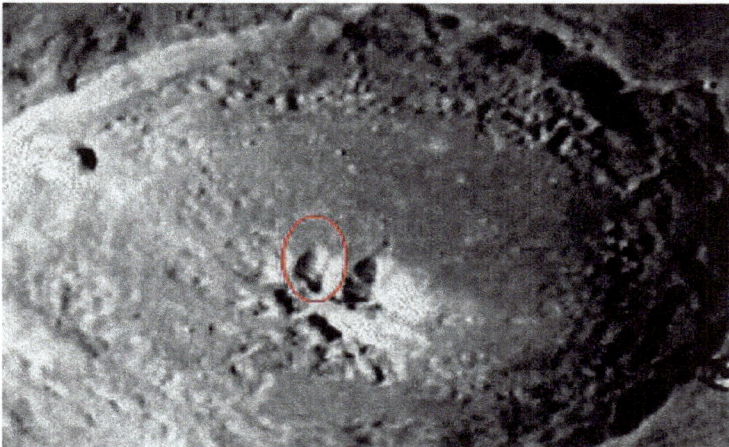

Below is a picture of a .38 cartridge to compare shapes. The purpose of the cylinder in the middle of the crater is unknown. Its size, however, must be huge when compared to the peak it is nestled against. And it just simply doesn't seem to belong here with the natural features. Does it?

Here is a view of the cylinder from another vantage point:

Closer:

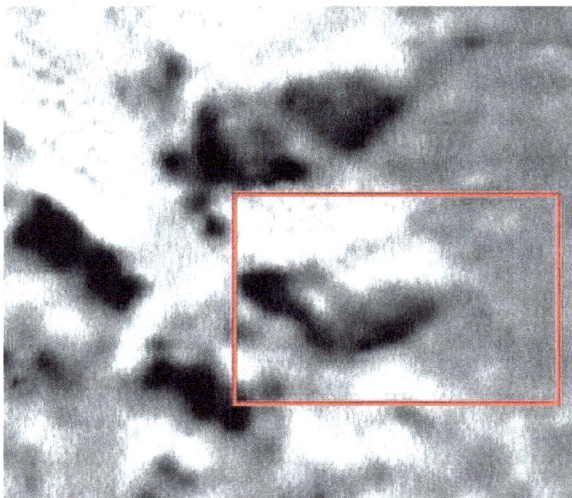

There's even more to see in this crater. It is on the side of the rim opposite the one where the "mesh" is located. Are all of those circular locations in the ground nuclear missile silos!

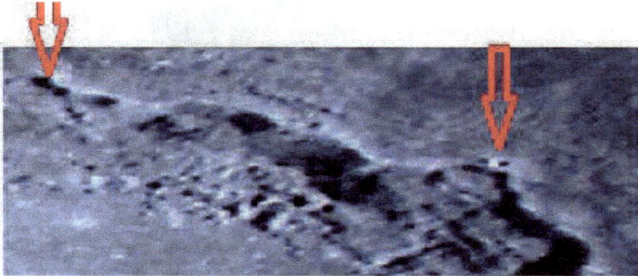

In the red square is what appears to be an Aztec like pyramid. Note the dark, flattened roof.

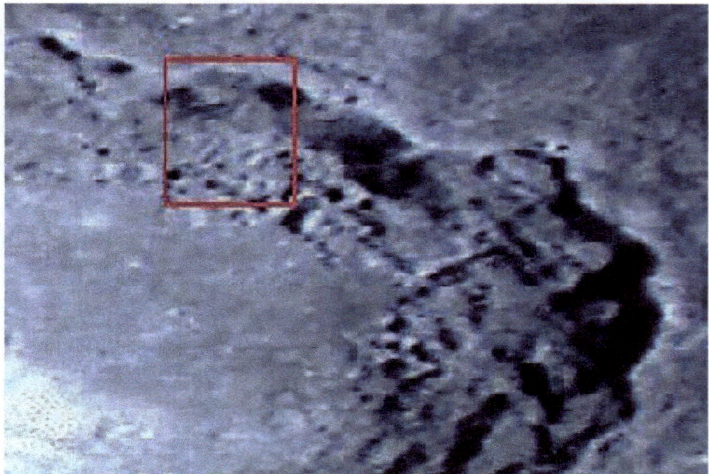

Aztec Pyramid

The gaps in the outer wall of this crater are artificially enclosed by what very much resemble the walls of a levee, some straight and some curved on top (see below).

Note below the part of the rim with the large curve in it that connects one portion of the "levee wall" with another just above the pyramid.

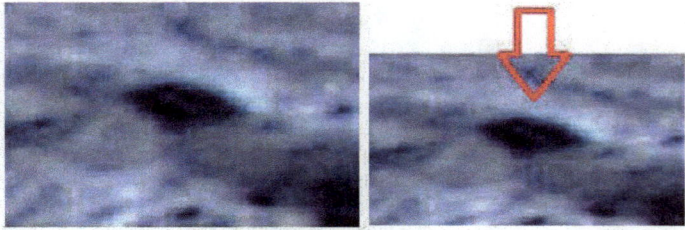

Next look very closely at a blow up of the above photo. Do not the red arrows on the left appear to be pointing at a series of very long stairs like those seen in Aztec pyramids? Does not the arrow coming up from the bottom appear to be pointing into a window of a building? Does not the arrow from above point directly at a curved artificial wall?

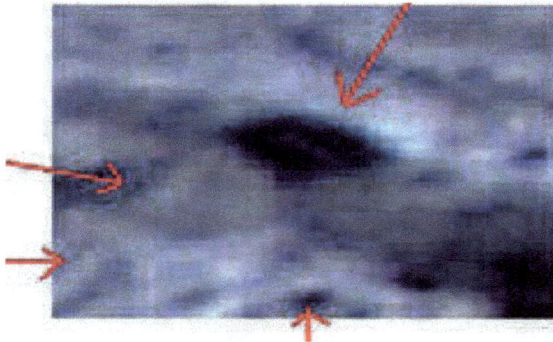

Now look at the section of the crater rim on the far right.

Notice the shiny circular shaped objects in the ground (possible missile silos?). And note the red wall at the top of this plaza. One can see the angular shape of the wall. It looks like red brick! Whatever it is made of it is not a natural formation.

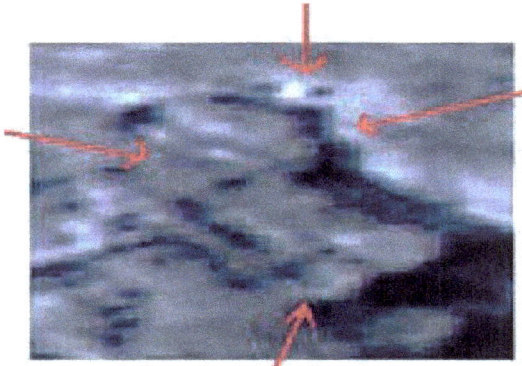

Does part of the right side of Crater Copernicus look like some type of religious sanctuary? It is possible that this portion of the crater was used by the ancient native inhabitants of the planet for mystical rituals. While at the same time the other side of the crater appears to be undergoing some form of construction by the current extraterrestrial rulers of this world. It isn't unusual for one civilization to appropriate the structures of another for its own use.

Remember the view of the center of the crater?

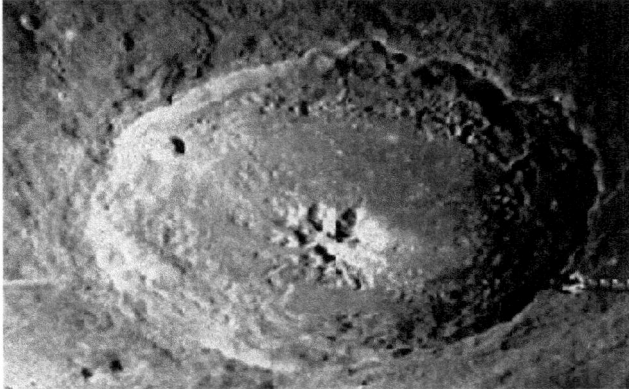

But according to NASA's photos, how can THIS below also be the center of the SAME crater?

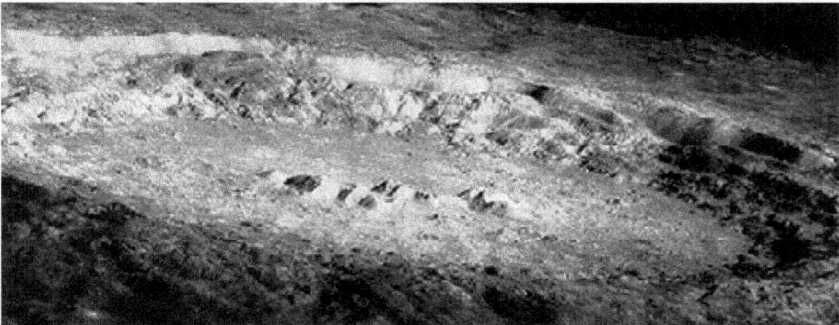

Close up:

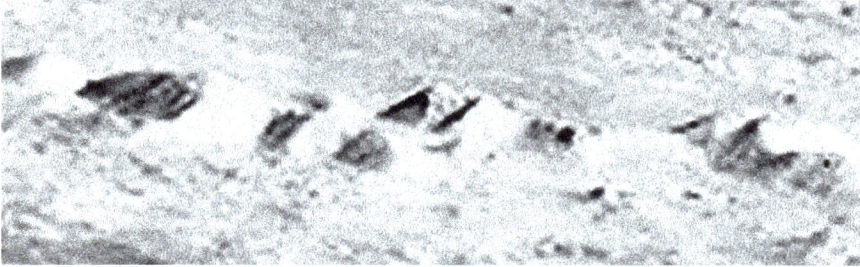

And what is this amid the peaks?

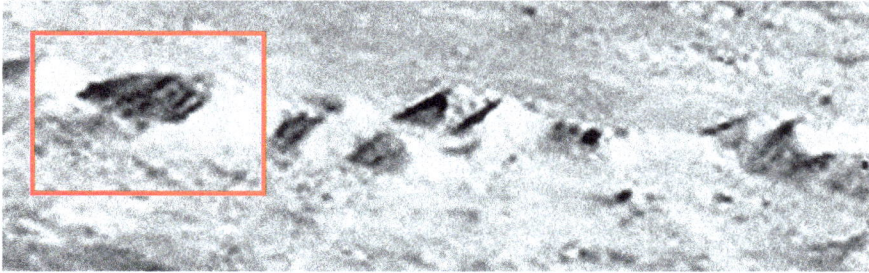

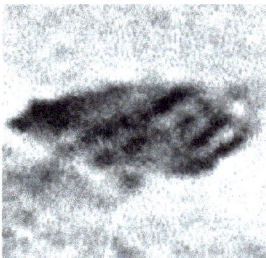

An alien spaceship buried in the dusty interior? Or a NASA commissioned **drawing** of an alien spaceship buried in the dusty interior?   Never underestimate the deceptive powers of NASA.

Here is another look at THIS part of the interior of the crater (in red).

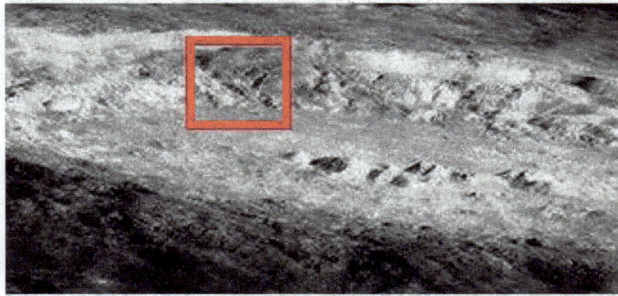

What is this?

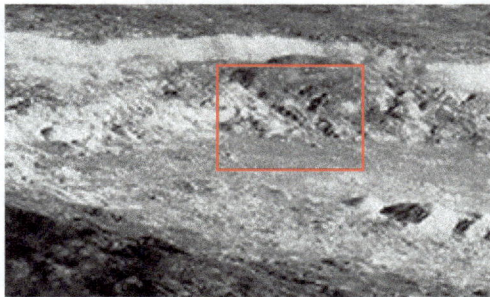

A closer view:

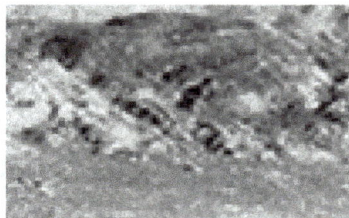

A drawing of portions of a crashed craft that had been thrown into the hillside? Or the real thing?

What's real and what's fiction? There can't be two different centers belonging to the same crater. After viewing hundreds of photographs of Copernicus dating back to 1903 it is clear that the picture of the crater with the grouping of pyramids and the apparent crashed UFO is a derivative and does not represent the true appearance of the interior of the crater.

Where did the pyramids come from, then? Maybe they were imported into the picture from here:

Why? Perhaps the purpose is to take attention away from all of the other anomalies which were just highlighted, including the woven mesh steel, the perforated metal, the levee like walls, the Aztec like pyramids and the shiny missile silo-like cylindrical objects located all along the hillsides. NASA would gladly exchange a grouping of mysterious Egyptian like pyramids and a drawing of a crashed UFO for that true collection of anomalies that exist in the genuine photograph of Copernicus!

Aliens also seem to be busy in the location of Mons Gruithuisen. Many people have noted the unusual dome shaped structure which seemed to have suffered a couple meteorite blows.

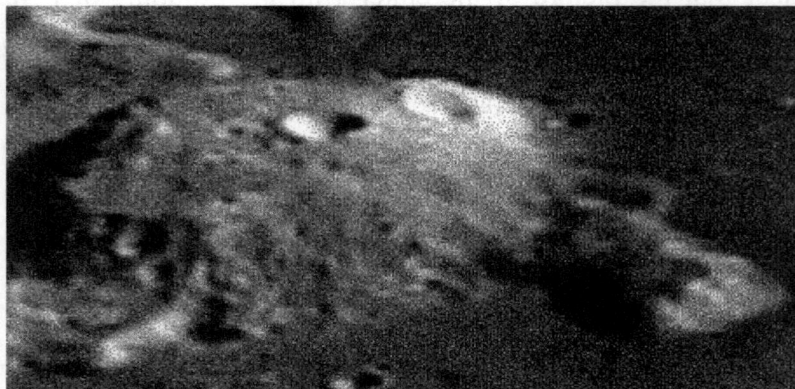

But not a lot of attention seems to be given to the newer area to the right of the dome. It is my belief that the old dome was abandoned after the meteoric impact and that operations were shifted to a more updated looking location to the right. See the picture below.

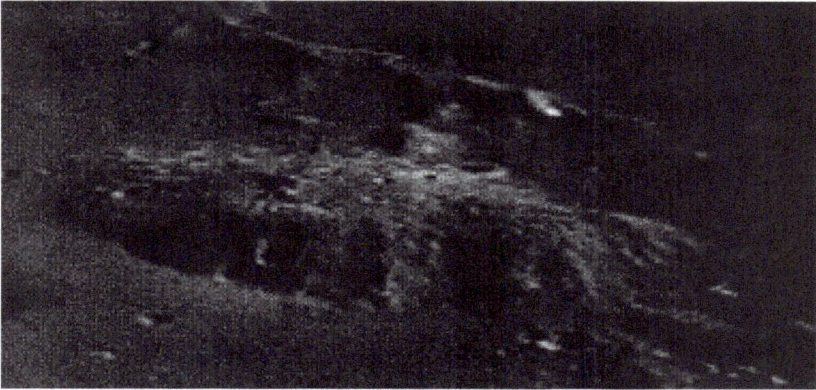

Notice the locations in this new dome at which the arrows are pointing. Do they not appear to be bay like openings cut into the walls? One of them seems to be open and the one on the far right appears to be closed.

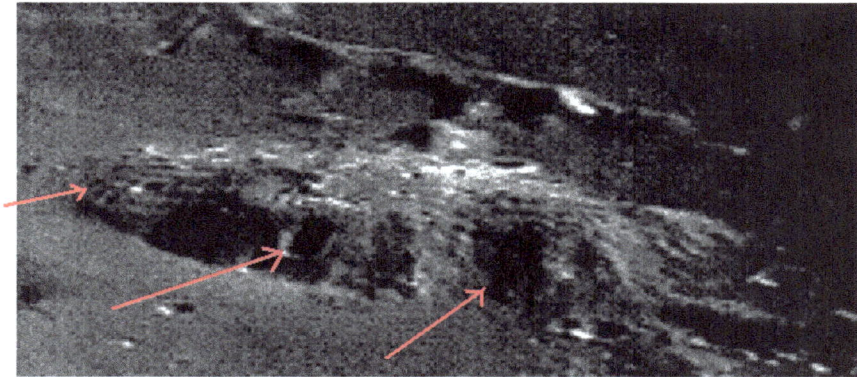

Note also the finely terraced walls on the left side of the dome. Is not this entire complex somewhat reminiscent of missile launching locations on earth as pictured next?

By the way, note the irony that in the above photo of an earth based missile launching site above there is a UFO in the upper left keeping surveillance on the location.

The home base of the UFO is almost certainly our Moon and it may even have a port in one of the bays in the southern portion of Gruithuisen which seems to have places ready to accept saucer shaped craft (as below).

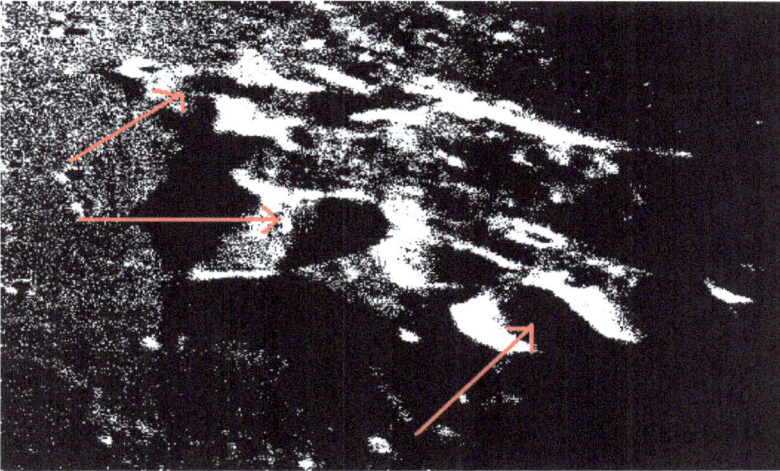

The arrows point to bays where saucer shaped vehicles may land. These are not naturally shaped craters.

This next object is almost impossible to just explain away as a peak in the center of a crater. Or is it? At first I suspected that it was a fictional drawing placed there by NASA to give anomaly hunters what they wanted – a grand anomaly. The crater in which the "object" appears is on the far side of the Moon and is named Tsilikovskii (various spellings are used). This is the anomaly below. (Be aware that this same anomaly was also once spotted in two other craters, including Crater Planck!)

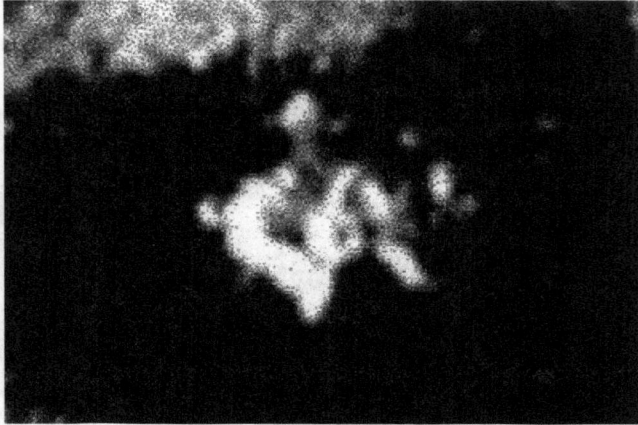

What it appears to be is some form of technological complex in the bottom center of the crater. It seems almost too amazing to be true.

We must wonder: how could this obvious anomaly have possibly escaped NASA's censors! Or is it created BY NASA.

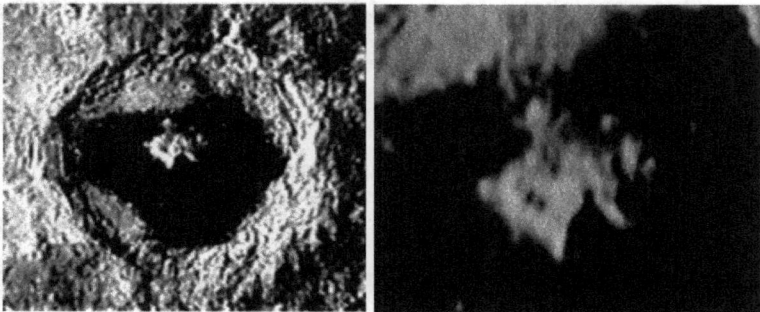

When a closer look is taken of the above photograph it was felt that the signs of editing could be seen in the white lines just above and just below the anomaly that had been added to Tsilikovskii. However, these lines appear throughout the entire photograph of this region of the Moon in the earliest photos (which this is) and, as noted, are common in the primitive type of photography then in use.

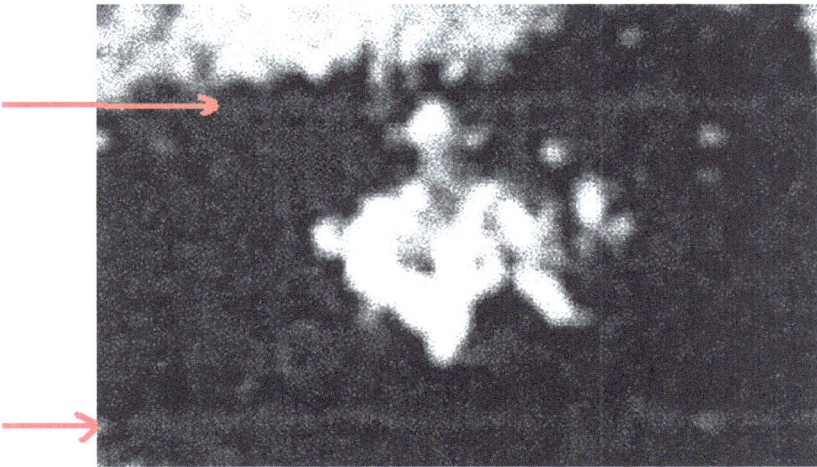

Also, the central tower that appears in this s c i e n c e building or tech complex is very similar to the one found in the complex in Mare Imbrium. This adds to the credibility of its truly existing in reality. Below is the suspected tower in the Imbrium complex.

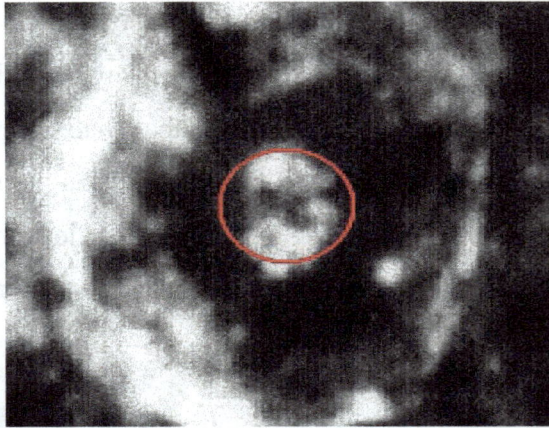

NASA is generally more interested in deleting real anomalies out of existence. So I chose a more recent photograph of Tsilikovskii to compare to the one which has the anomaly. The two photographs that follow were taken by Apollo astronauts from orbit while the earlier photographs were taken by lunar orbiters.

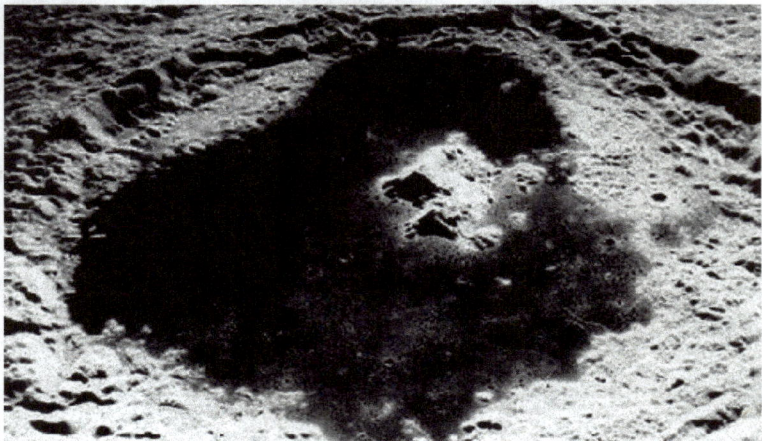

Next is an extreme close up of the interior.

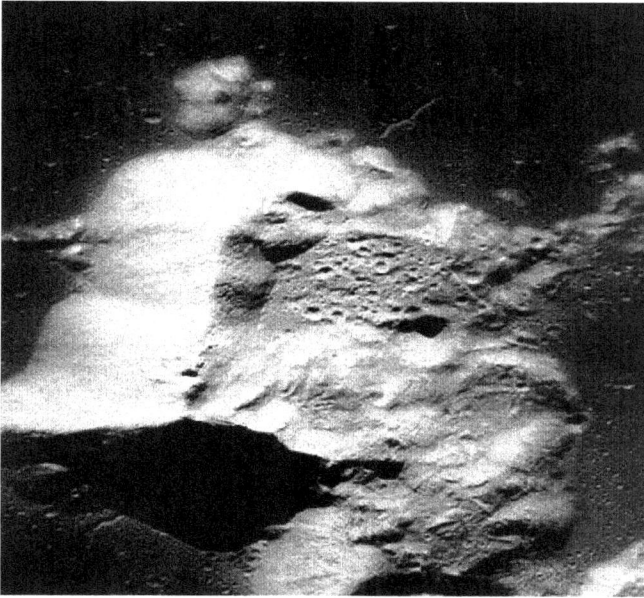

Compare the old version of the interior with the new version:

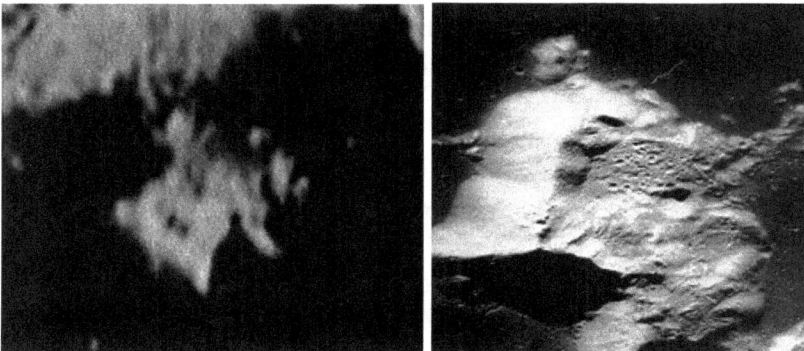

There doesn't seem to be much similarity.   This appears to be a form of NASA trickery.   Either NASA edited the "anomaly" into the earlier photograph or edited it OUT of the newer photograph. There cannot be two different centers of a crater existing at the same time as already noted in Copernicus. Basic physics. Which do you think is more likely? That NASA created an anomaly for people to wonder about? Or did they replace an anomaly to keep it secret? More data is needed for a definitive answer. By the way – remember – this **same** "anomaly" also appears in two other craters on the moon.

Sometimes, however, the person at NASA in charge of the deleting process gets confused. He was very confused when it came to the case of the Crater Lamb. Note the obvious black bar at the bottom on the rim of the crater shown below. NASA isn't very subtle when it doesn't want the taxpayer to see something. They just block all view of it.

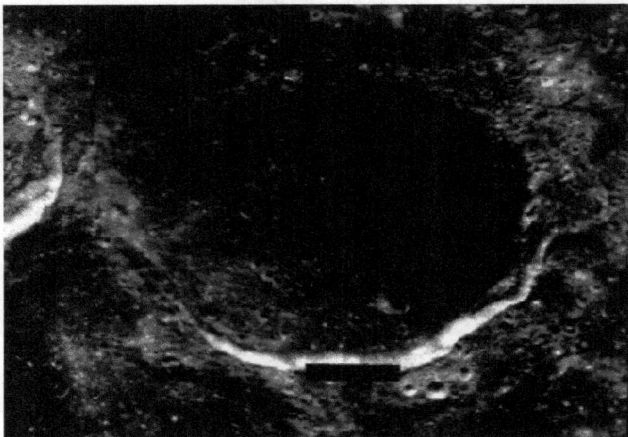

Apparently, it was the top rim of the crater that should have been blocked out and not the bottom. The censor got it wrong.

Remember this image of a tube-shaped UFO that had landed in Mare Imbrium:

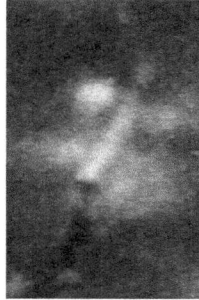

It turns out that the same type of UFO was also photographed on the TOP rim of Crater Lamb. It isn't known whether or not it's a picture of the same UFO but it certainly is the same type of UFO. Below is what was photographed on the top rim of Lamb crater:

Ironically, a photograph of Lamb crater before it was censored was located and below are two pictures of what is on the lower rim which we weren't supposed to see:

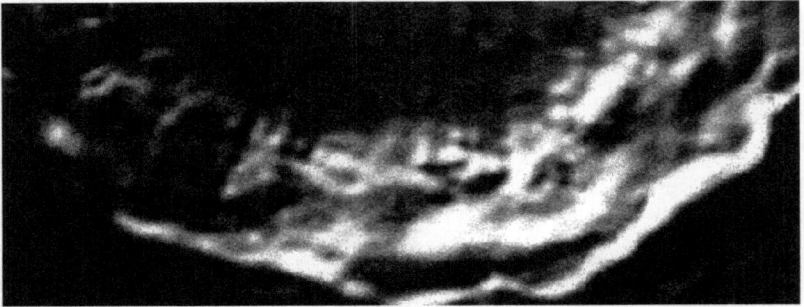

And closer...

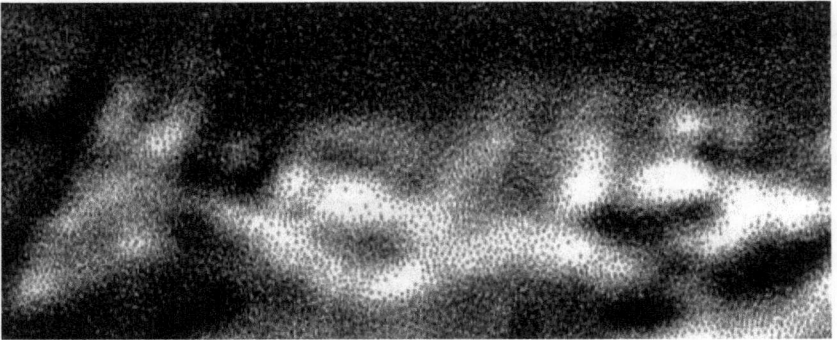

There does not seem to be anything too incriminating here. I guess NASA didn't know which rim the UFO was on. Well, it is a large Moon to try to keep hidden from view, isn't it?

But that isn't the end of this matter. In the NASA censored image above almost no one notices the area in the center of the crater that is also blacked out. This is because the area is already covered in shade. The Crater Lamb will be shown again but this time with the blacked out portion in the center noted by a square red box. Then we'll see what else NASA was trying to keep us from seeing. There is actually a black stripe in the middle of the crater placed there by NASA. The red arrow points directly to its edge.

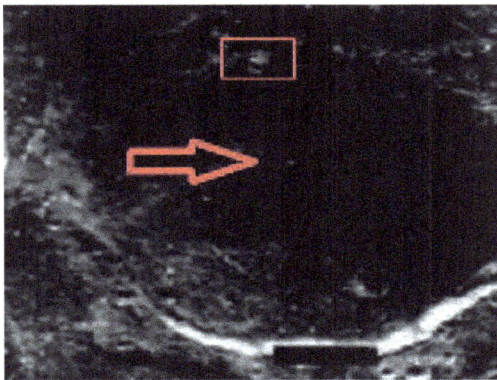

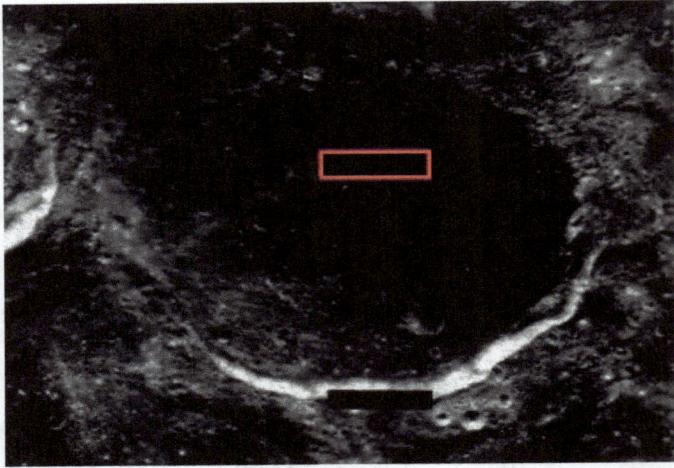

Here is a look at the same crater uncensored and with a full view (slightly turned):

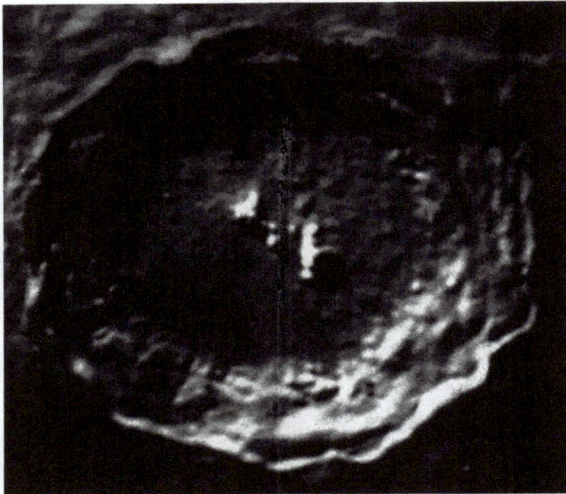

NASA can't get this one right. On the top rim, they forgot to white out the cigar shaped UFO. But in the very center of the

original photo they forgot to block from view an alien craft that was seen once before in Mare Imbrium. Remember the following:

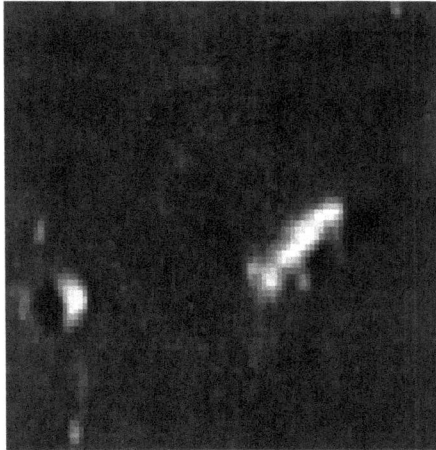

Isn't this the same device that is now in crater Lamb? Or is it a duplicate device?

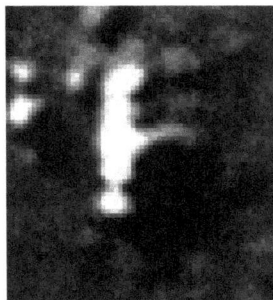

Lamb

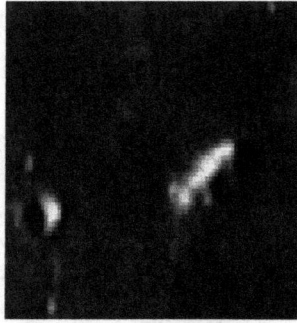

Imbrium

And the same type of legged UFO was also seen at both sites. Yes, it is indeed a large Moon to try to keep hidden from view, isn't it?

Encounters between Astronauts and Aliens

In this section the truth of the secret discussions between the astronauts on the Moon and NASA will be revealed. While the astronauts were involved in their coded conversations, photographs were being taken from the orbiter above, showing what the astronauts on the surface were describing. It is visual evidence that these conversations did take place. The images will follow below.

Encounters between astronauts and aliens were almost common upon the Moon. They started with the very first mission, Apollo 11. Not long after stepping onto the lunar surface, Neil Armstrong took the following photograph:

These are a pair of UFOs that he reported seeing to Houston.

After brief conversation during the early parts of the first stroll on the surface with Buzz Aldrin, the communication with earth seemed to suddenly be cut off. Communication wasn't really cut off but it was switched to a different mode, a coded system which prevented the people who'd paid for this mission – the US taxpayers – from hearing and seeing what was really transpiring.

However, the verbal part of the communication was captured by many amateur radio operators on earth who were able to tap into the secret system being used by NASA. And the conversation that they heard between the astronauts on the Moon and mission control was pretty shocking. This type of communication interception by amateur radio operators occurred throughout the Apollo Program. It is the purpose of this section to **SHOW** you what the astronauts are referring to when discussing anomalies with Houston. Fortunately, this author has most of the original photographs of the anomalies that were discussed and which NASA later thoroughly expunged.

This examination starts with Apollo 11 and the photograph above taken by Neil Armstrong. What follows is what was being said among Armstrong, Aldrin and Mission Control about the alien craft that was performing surveillance upon them.

Astronauts (either may be speaking): What was it? What the hell was it? That's all I want to know.

Houston: What's there? Mission Control calling Apollo 11.

Astronauts: These babies were huge, sir! Enormous! Oh, God! You wouldn't believe it. I'm telling you there are other spacecraft out there…They're on the Moon watching us.

Armstrong and Aldrin move onward a little and then are once again shocked by more UFO.

Astronauts: Those are giant things. No, no, no – this is not an optical illusion. No one is going to believe this.

Houston: What…what…what! What the hell is happening? What's wrong with you?

Astronauts (unclear which is speaking): We saw some visitors. They were here for a while observing the instruments.

Houston: Repeat your last information!

Astronauts: I say that there were other spaceships. They're lined up in the other side of the craters.

Houston: Repeat. Repeat.

Astronauts: My hands are shaking so badly I can't do anything. Film it? God, if these damned cameras have picked up anything – what then?

Houston: Have you picked up anything?

Astronauts: I didn't have any film at hand. Three shots of the saucers or whatever they were that were ruining the film.

Houston: Control, control here. Are you on your way? What is the uproar with the UFOs over?

Astronauts: They've landed here. There they are and they are watching us.

Houston: The mirrors, the mirrors – have you set them up?

Astronauts: Yes, they're in the right place. But whoever made those spaceships surely can come tomorrow and remove them. Over and out.

Later after returning to earth, Armstrong had a lot to say during a private conversation that very few people have heard about. He told a friend that we were warned to stay away from the Moon. Don't plan to set up any stations or begin work on a permanent city. He further said that we completed most of our already planned missions and were not considering returning.

APOLLO 16

Apollo 16's landing site was in the Descartes Region of the Moon. It's a very hilly boulder strewn terrain which quite impressed Charles Duke as he strolled across it.

Duke: YOWEE! Man, John I tell you that this is some sight here. Tony, the blocks in Buster (crater are covered – the bottom is covered with blocks five meters across. Besides, the blocks seem to be in a preferred orientation, northeast to southwest.

Below is a photo of the Crater Buster. It is the approved NASA version.

They then climb what is known as Stone Mountain and Astronaut Duke is really impressed here. It is pictured below – the approved NASA version.

Duke: If this place had air it would be beautiful. It's beautiful with or without air. The scenery up on top of Stone Mountain, you'd have to be there to believe it – those domes are incredible.

Houston: O.K. could you take a look at that smoky area there and see what you can see on the face?

Duke: Beyond the domes the structure goes almost into the ravine and you can't see the delineation. To the northeast there are tunnels, to the north they are dipping east to about 30 degrees.

Domes, a structure of some type, and tunnels. Sounds like artificial structures don't they? And since Duke was so impressed by the sight, there must be lots of pictures of them. Unfortunately, not. At least, not from lunar ground level. But NASA was tricky and, instead of showing us the pictures that Duke certainly took, they decided to offer a realistic drawing.

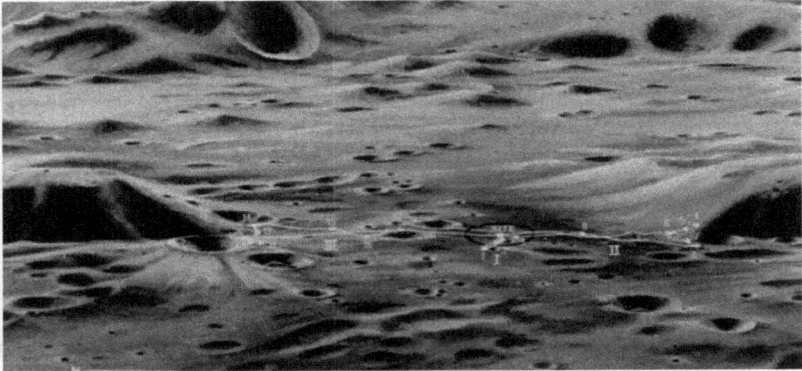

However, other people can be just as tricky as NASA. Below are photographs taken from orbit of the same area. Note the tunnels in the close up view.

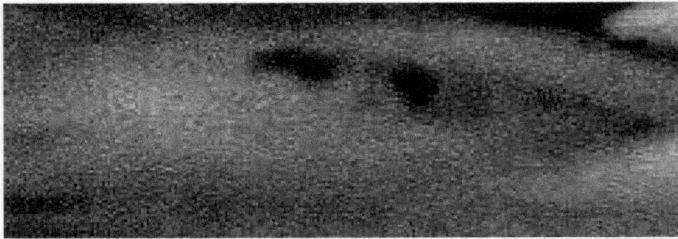

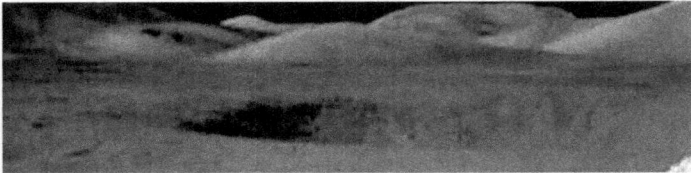

Sorry we couldn't get any closer. But these are the tunnels burrowed into the lunar hillsides by which Duke was so impressed. Behind and beside them are the domes he was so raving about.

A glimpse was able to be caught of that "structure" he referred to dipping into a ravine. You'll have to look close, but this appears to be it:

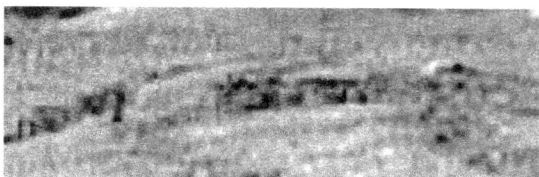

Another view of what appears to be a stone structure of some type:

For clearer pictures, we'll need to ask NASA. After all, they've been paid for already.

You probably didn't notice anything out of the ordinary in the photograph above of Stone Mountain. What was the astronaut so worked up about, anyway? Whatever it was, NASA made sure we didn't see it. The photo has been well airbrushed, making Astronaut Duke look like some kind of idiot who is raving about this perfectly ordinary looking location on the Moon. Unfortunately, I haven't been able to uncover any incriminating photography in this instance. But the attempted cover up of the real landscape around the Apollo 15 landing site is another matter. Sometimes NASA cannot edit every potential piece of evidence.

APOLLO 15

First, a photograph of one of the Apollo 15 astronauts under the observation of a UFO in the background.

Many researchers concerned with lunar anomalies have studied the infamous Mount Hadley cover up. Apollo 15 landed in the Apennine Mountains and the astronauts involved were David Scott, Alfred Worden, and James Irwin.

We begin with the astronauts' own words in a coded conversation that once again was saved for us by the true heroes of the airwaves – the independent radio operators. This time, instead of raving about terraces on the side of Stone Mountain, these astronauts are raving about the amazing patterns and tracks of some type that were found around Mount Hadley.

Scott: Arrowhead really runs east to west.

Houston: Roger, we copy.

Irwin: Tracks here as we go down slope.

Houston: Just follow the tracks, huh?

Irwin: Right, we're (garbled). We know that's a fairly good run. We're bearing 320, hitting range for 413. I can't get over those lineations, that layering on Mt. Hadley.

Scott: I can't either. That's really spectacular.

Irwin: They sure look beautiful.

Scott: Talk about organization.

Irwin: That's the most organized structure I've ever seen.

Scott: It's so uniform in width.

Irwin: Nothing we've seen before this has shown such uniform thickness from the top of the tracks to the bottom.

Wow! Sounds like something really amazing. Is this it below?

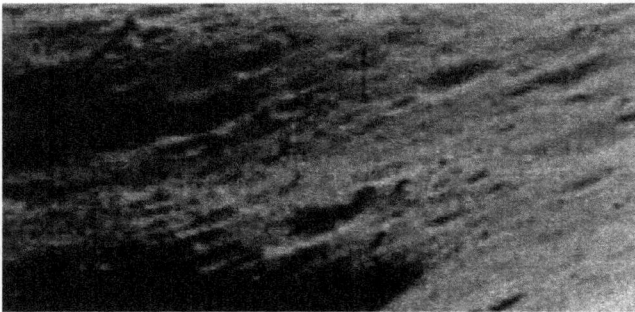

As you can see, above is a composite picture that has been heavily airbrushed. It's pretty obvious where this picture was pieced together from other photographs. Once again our astronauts look like raving lunatics. What is so awesome about Mount Hadley! Where are these marvelous patterns and these rows of tracks? And the lineation!

Here they are! Below is a photograph of the lineation the astronauts spoke so much about but which weren't shown in their own photos:

Below: NASA censored beside actual photo.

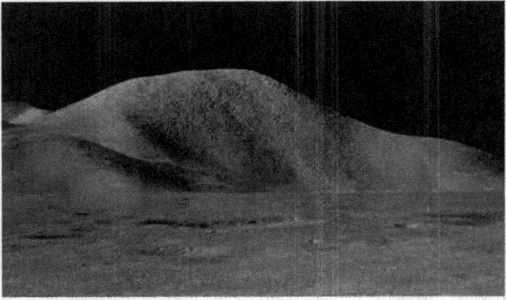

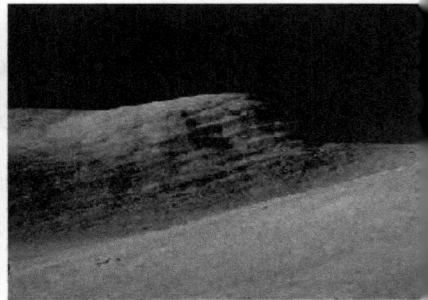

Below are the tracks that were also repeatedly mentioned. Apparently, the tracks are so impressive simply because they weren't supposed to be there. They show signs of other activity on the Moon, kind of like what the Apollo astronauts' own tracks look like. These were already impressed into the surface before the astronauts arrived. Made by alien astronauts?

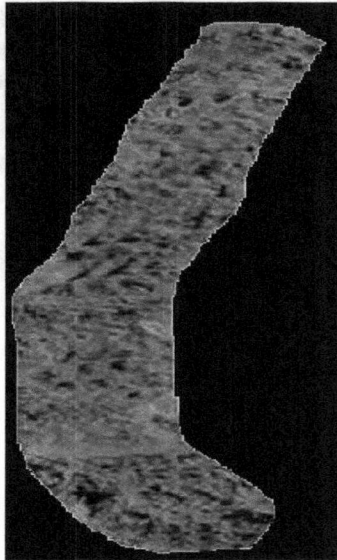

These are the markings in the surface that were airbrushed out of the "original" photograph of Mount Hadley. They must have been even more impressive in person.

Apollo 17

The Apollo 17 crew was made up of Eugene Cernan, Harrison Schmidt, and Ronald Evans. It operated in the Taurus Littrow Valley region of the Moon. It is this mission which produced one of the most remarkable pieces of evidence of artificial artifacts on the Moon. This book shows the first pictures of what they found. And it is pretty amazing. First, their discussion of the matter.

Houston: Go ahead, Ron.

Evans: Roger, I guess the first thing I want to report from the back side is that I took another look at the – the – cloverleaf in Aitken (crater) with the binocs. And that southern dome (garbled) to the east.

Houston: We copy that, Ron. Is there any difference in the color of the dome and the Mare (?) Aitken there? (Note: Aitken is a crater. I assume he means the crater floor).

Evans: Yes, there is. That Condor, Condorsey, or Condorecet or whatever you want to call it there. Condorecet Hotel is the only one that has got the diamond shaped fill down in the, uh – floor.

Houston: Roger. Understand. Condorcet Hotel.

The following photograph may be referring to this weird "hotel" which appears to be some type of gigantic artificial structure. This photo was taken from orbit, looking directly down onto the structure. This is the first identification of it with the astronauts' report just given. Note the pattern on the "floor" to the middle right side of the picture.

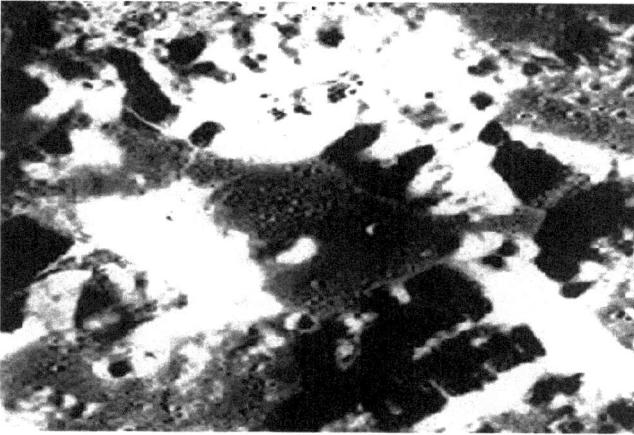

Since the view is from above how can there be areas of shadows without anything casting them? This implies openings in the "structure" rather than shadows.

The picture of the Crater Aitken is next. And this is where a shocking revelation is found. I doubt NASA was aware of it or they would have kept this view from being made public at all cost. Again, the picture is taken from orbit. First Aitken will be shown from high above.

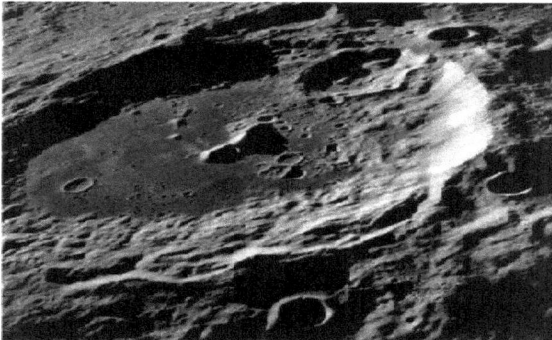

Nothing too strange, is there? It looks like a typical non-controversial lunar crater. In fact, it looks quite average. Almost textbook. But let's get a little closer.

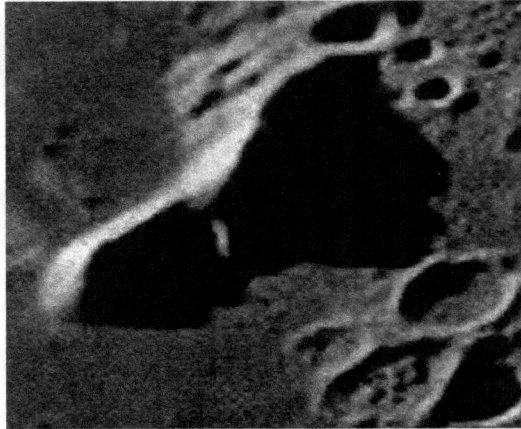

Now things begin to get interesting. Do you notice it? It isn't an optical illusion and it isn't an artifact of the photography. It is there. In fact, it is **EXACTLY** as described by astronaut Evans. He was confused about its location (or purposefully confusing), but not about what it looked like: "…is the only one that has got the diamond shaped fill down in the, uh – floor." Also, one can see the "cloverleaf" pattern he mentioned. Notice enclosed in red below:

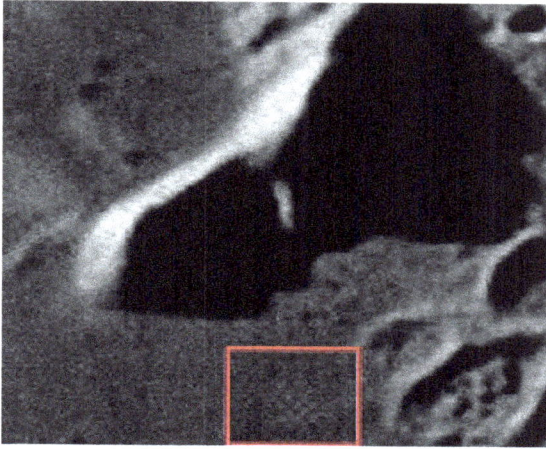

An even closer view:

There is clearly something of artificial design in the floor of the crater Aitken! Too bad NASA didn't tell us anything about it. It's not the same as the material already discovered in Copernicus. But we still don't know what this is, either. Could it be the remnants of some form of floor tile? Diamond or cloverleaf shaped? That IS what it looks like. Floor tile!!

Apollo 17 seemed to be on the lookout for alien or other type of intelligent activity taking place on the Moon. Read the following supposedly secret conversation just to see how involved with this they were.

LMP (Lunar Module Pilot): Where are your big anomalies? Can you summarize them quickly?

Capcom: Jack, we'll get that for you on the next pass.

CMP (Command Module Pilot): Hey, I can see a bright spot down there at the landing site where they might have blown off some of the halo stuff.

Capcom: Roger. Interesting. Very – go to KILO. KILO.

CMP: Hey, it's grey now and the number one extends.

Capcom: Roger: We got it. And we copy that it's all on the way down there. Go to KILO. KILO on that.

CMP: Mode is going to HM. Recorder is off. Lose a little communication there, huh? Okay, there's bravo. Bravo, select OMNI. Hey, you know you'll never believe it. I'm right over the edge of Orientale. I just looked down and saw the light flash again.

Capcom: Roger. Understand.

CMP: Right at the end of the rill.

Capcom: Any chance of -?

CMP: That's on the east of Orientale.

Capcom: You don't suppose it could be Vostok?

Vostok was a Russian probe. None of them ever went to the Moon here. No, it couldn't be Vostok.

The words speak for themselves. It probably was an alien craft that Apollo 17 spotted sending signals from the edge of the Orientale. Possibly the very same UFO that later became airborne and was observing the Apollo 17 landing site from above as in the photograph below. Rather ironic, isn't it?

This concludes a frustrating section in which NASA spends most of its time hiding from its taxpayers what its astronauts are reporting. Note the words: "Lose a little communication there, huh?" And: "Recorder is off." No, they weren't bothered at all about keeping their findings secret from us, the ones who sent them there to tell us what they found!

Settlements of the Ancient Selenites

Selenites were what the original inhabitants of the Moon used to be called when it was the common belief that the Moon was inhabited. The word Selenite comes from the ancient Greek word for the Moon, Selene. I'm going to continue using the word Selenite because there aren't any other good alternatives and in reality it is quite appropriate.

Belief that the Moon was inhabited continued well into the twentieth century. And it wasn't only people with vivid imaginations who held this belief but brilliant men of science. Two of the most well-known astronomers who ever lived – John and William Herschel – both accepted the possibility of life existing on the Moon.

One of the most renowned astronomers of the 20th century, a man named William Henry Pickering, was in charge of creating the first true lunar atlas in 1903 based on astounding telescopic photographs. He often reports on cases where snow and clouds were found in and around various craters, implying the existence of a limited atmosphere.

"The rim of this crater (Pallas) is here seen to be white with snow. In plate 9C a large part of this whiteness has disappeared, and in 9D and 9E the rest of it has vanished."

Below is a modern photo of Pallas in the Sinus Medii region of the Moon. It is in the red square. Compare Plates 9C, 9D and 9E from the 1903 lunar atlas for a view of the melting snow on the rim of Crater Pallas.

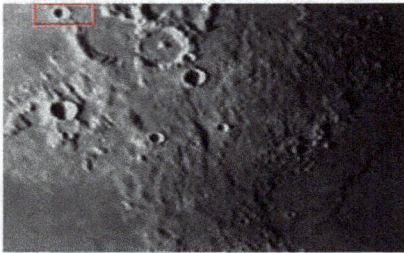

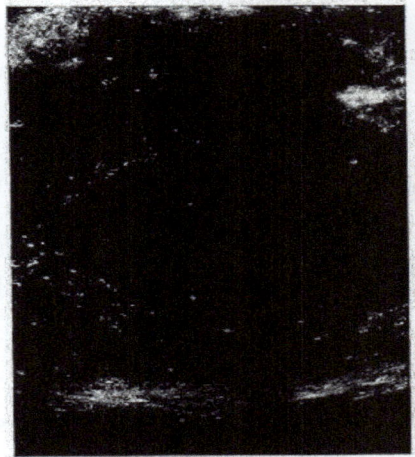

PLATE 9C

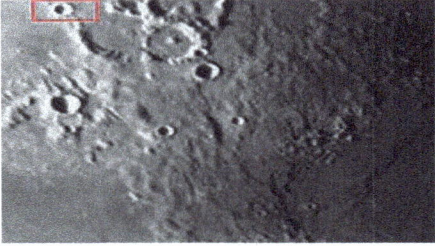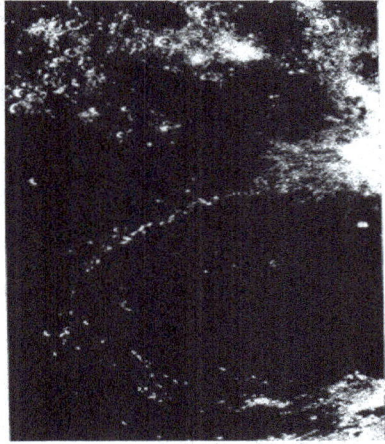

PLATE 9D

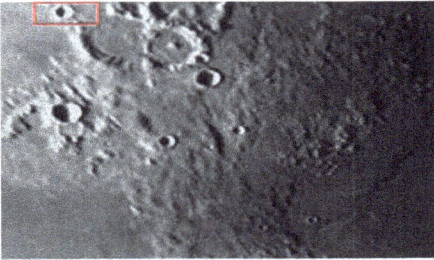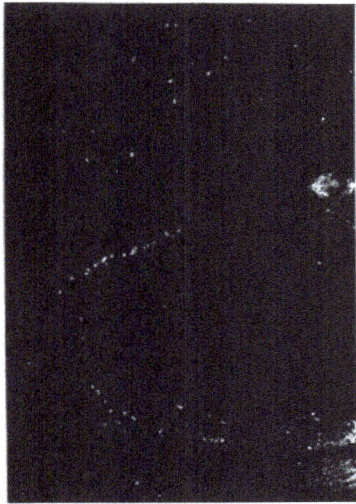

PLATE 9E

"It was shown in these Annals XXXII, 207, that the white spot surrounding the interesting crater Linne…is subject to fluctuations in size analogous to those of the polar caps on Earth and Mars." Linne is below.

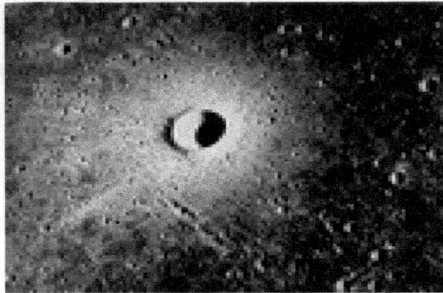

This is to suggest that there still exists even today a much rarified atmosphere on the Moon and even areas of water. The existence of both has been proved by both the Apollo missions and later investigations.

But the only living beings on the Moon currently are the extraterrestrials. We have already seen some of their structures but most of their complexes are probably underground.

The photographs you are about to see are concerned with the ancient, original civilization and the flora and fauna that occupied this world perhaps a billion years ago. A billion! How could any trace of them at all remain? When there is lack of a true atmosphere, decay and erosion do not occur. That is how examples of the remains of extremely ancient cities and deceased animal life forms can still exist.

It is generally accepted that the Moon was created from the earth. During the period of the earth's formation our world was struck by a massive asteroid. The debris from this strike was hurled into orbit around the earth. The debris then coalesced over the millennium to form our satellite. The samples returned to earth from the Apollo missions prove that the rocks on the Moon are made up of the same material that makes up the earth. There is one huge problem, however. The lunar rocks are almost completely devoid of water unlike the material on earth. This discrepancy has plagued scientists for decades.

While it seems likely that the water had been removed from the lunar material by baking in the intense heat of the sun over the millennia there is another possibility that would probably seem too farfetched for the standard astronomer. Could it be possible that the original inhabitants of the Moon found a way to siphon every last drop of water from the lunar material as they desperately fought to survive? Or could it alternatively be possible that the extraterrestrial visitors have done that same thing with their highly sophisticated technology? Maybe there is some truth in both ideas.

If there was an original civilization on the Moon how did it develop? It could have evolved the same way that civilization on earth did. It's possible that the Moon at one time had the same type of climate as on earth. But it doesn't seem there would be enough time for a civilization to develop on the Moon because it lost 99 percent of its atmosphere relatively quickly. It's simply a mystery for now. Maybe they did not need the same environment as we on earth do to survive!

It is also likely that flora and fauna were native to the Moon and were probably of gigantic size. This is because of the lesser pull of gravity on the Moon which would likely produce fauna of a large enough size to be able to navigate the land in a "normal" manner. Many remains of what resemble giant creatures have been photographed, most of them in the bottoms of deep craters. Why there? Because in a world that is rapidly being depleted of oxygen the bottom of a crater would be the final place to search for the last drop of air. Still a mystery.

Most people have seen what a spider that had just been stepped on forms into – a curled up dried out form. Is this some type of arachnid like creature in the center of the crater above? Notice the top rim and how a pair of spidery arms seem to be clawing to get out. Of course, it is no longer alive and will never escape the crater.

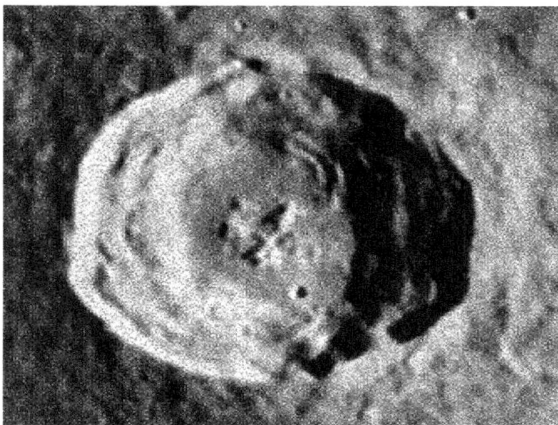

Another crater with some kind of bugs in its bottom?

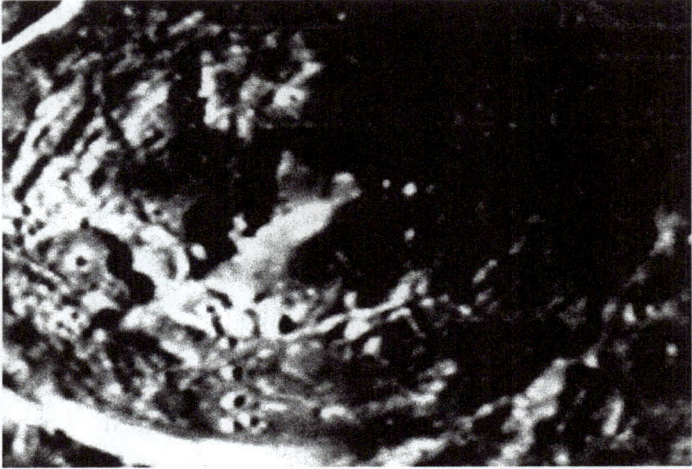

This crater holds something like a bat or a bird in various stages of decomposition. Note what appears to be a wing on the left side (in red below). This is in Crater Doppelmayer by the way.

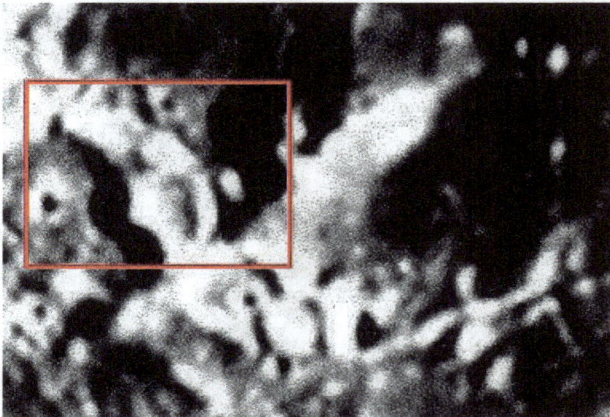

A caterpillar or lunar worm in the next crater?

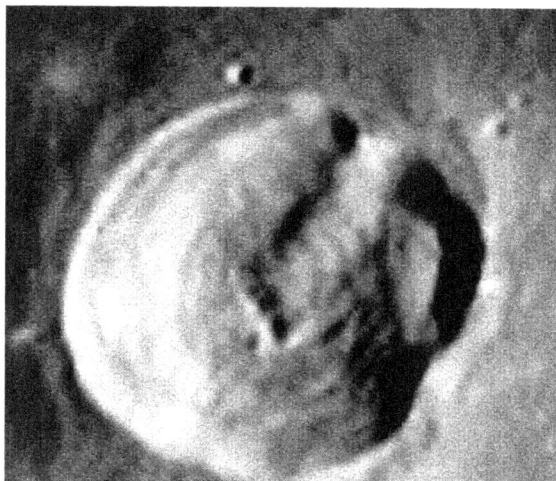

This is just a brief sample of a much larger menagerie of deceased creatures that some believe fill many of the craters on the Moon. It is very ironic that science fiction writers such as H.G. Wells (First Men in the Moon) described the Moon as being inhabited by oversized insect like beings. Is this something a lot more than just coincidence?

What cannot be mere coincidence are the similarities between the photographs of ancient deserted cities on the Moon and similar archaeological sites on earth. My specific training is in archaeology so I have an extensive background in this area. Thus, if I saw a site like that below on earth I would automatically consider it to be an abandoned ancient location.

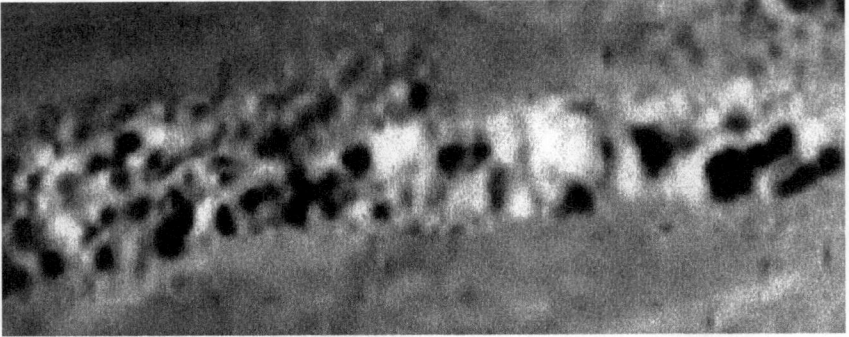

But this is on the Moon, not the earth. It is a collection of ancient stone buildings that lie in the shadow of the crater Eratosthenes.

A wider view.

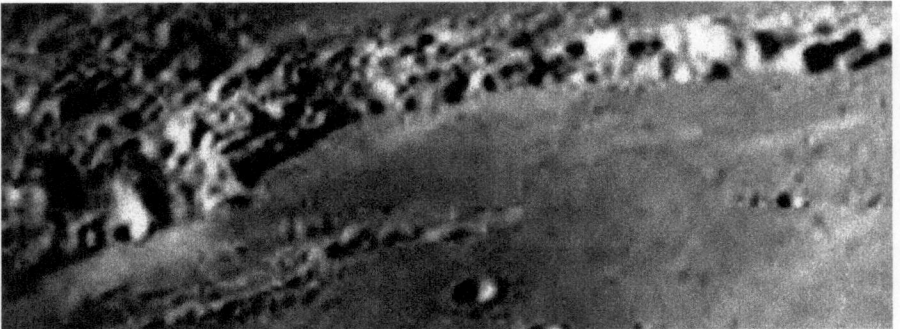

Above is an ancient lunar city. Below is an ancient Incan earth city.

They are very similar. But of extreme importance is what lay at the entrance of both cities. It isn't a coincidence. This is the buffer zone which helps prevent unwanted intrusion into the city. These buffer zones are the areas enclosed in red.

Incan

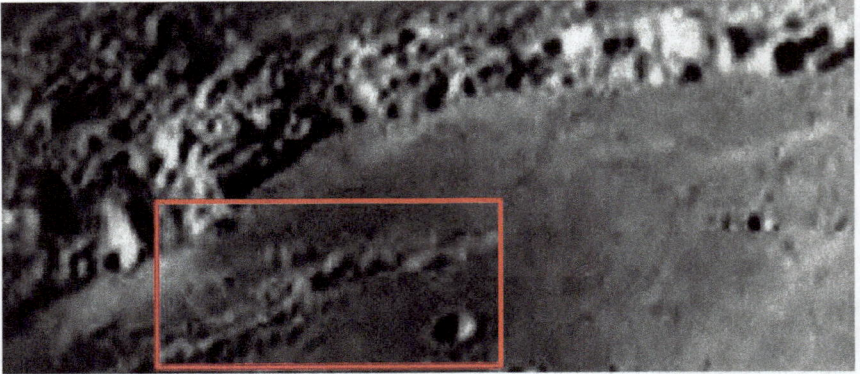

Lunar

A closer look at the lunar city is in order, starting at the partial building that is closest to the crater. It is pictured below. Can this be anything other than a former building?

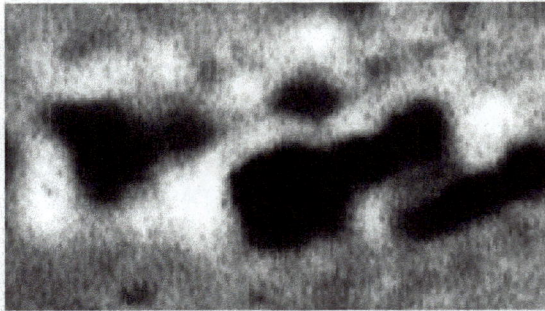

This is clearly not simply a jumble of fallen crater material. It has the outline of the walls of a former building just like the ruins of ancient earth Bronze Age cities like the one below.

Are these not almost identical?   See them side by side and judge.

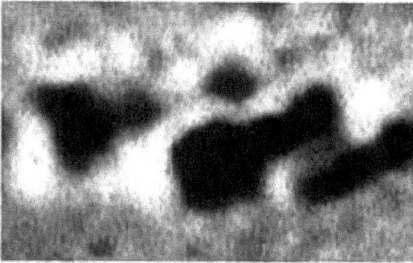   

Lunar   Bronze Age

Going toward the left within the lunar city it appears to be a typical Bronze Age village that has been worn by time or perhaps ruined by warfare. It is not of alien design, but belonged to the original lunar native inhabitants - Selenites.

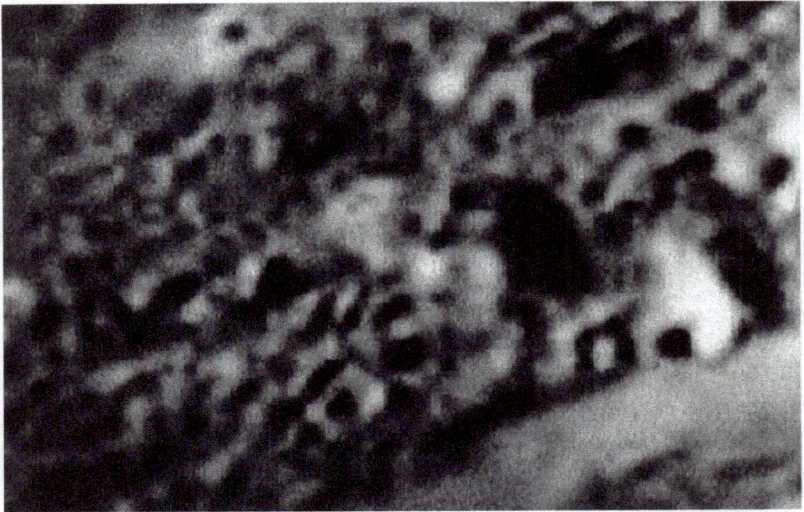

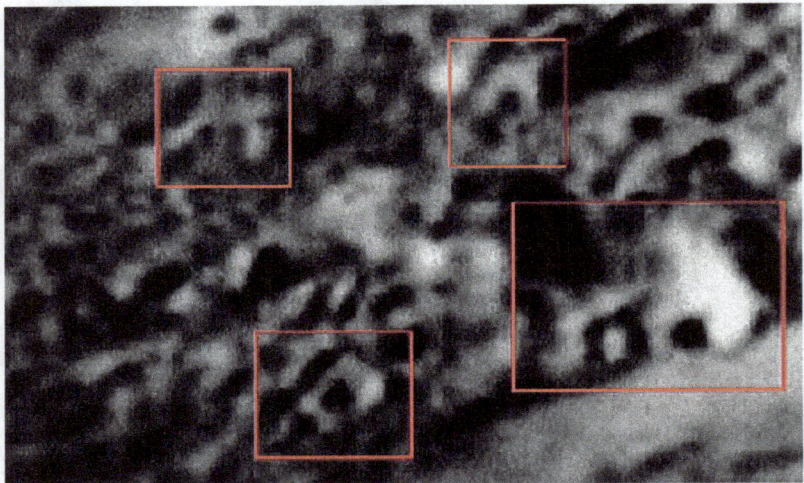

Note the recurring house like structures within the red squares. But most especially note the one on the bottom right. How can this be anything but a house and another structure attached to the left? A close up makes this even clearer.

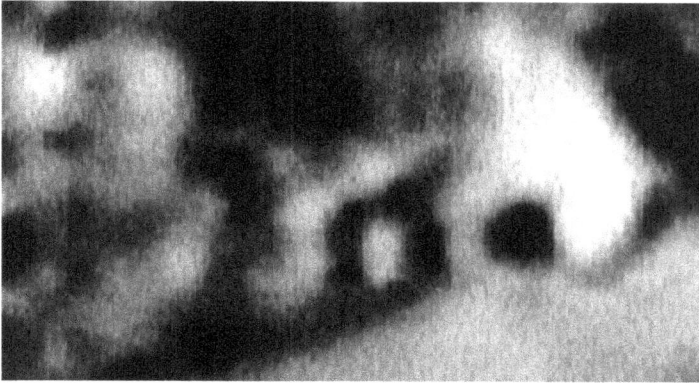

What makes this structure even more amazing is that there is a similar ruin like it in an ancient Bronze Age site on earth! See it below.

And like many of the ancient cities of earth this ancient lunar ruin was associated with some type of great edifice like a castle or a coliseum or cathedral. In this case the edifice appears to be a castle. It was constructed within the crater itself and is examined in the close up photographs that follow.

First is the photograph of the Crater Eratosthenes which contains the structures about to be studied. Keep in mind that even in modern times activity of unusual nature has been seen occurring within the Crater Eratosthenes. William Pickering associated it with the possible migration of various species of animals. The most likely cause of the movement seen in Eratosthenes would be alien archaeologists conducting excavations of this fascinating site. Too farfetched to be believable? Take a close look at the upcoming photographs.

It is almost impossible to look at this picture and NOT see the many buildings which fill the inside of it.

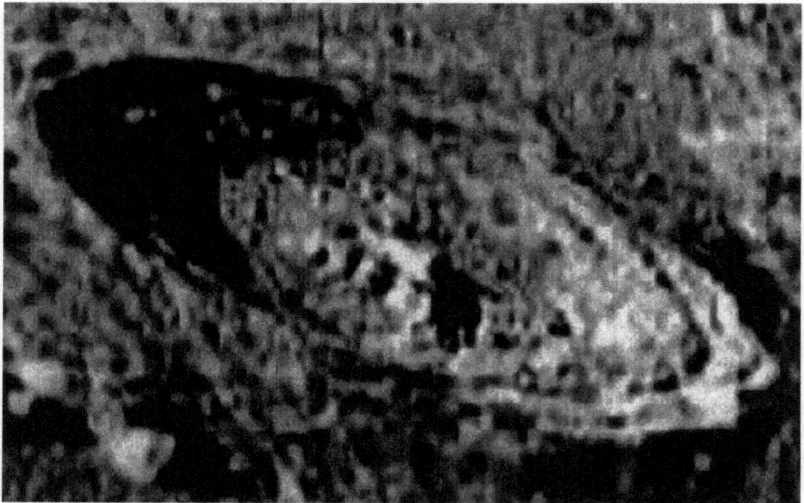

Next the focus is on the left side of the crater.

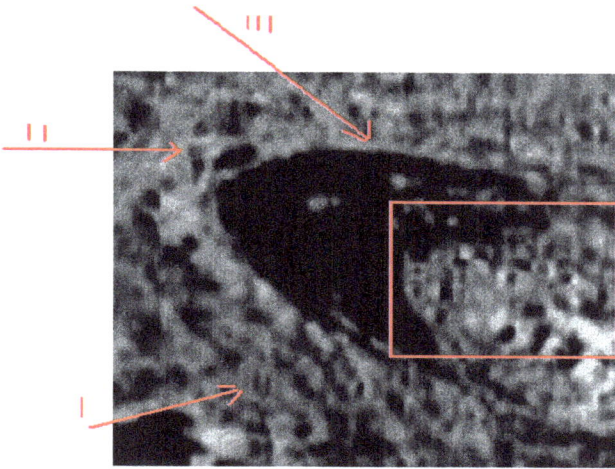

In the above photo the red I points toward a pillared area. The red II shows the backing of the structure at its highest part. The red III shows the perfectly sloping curved wall. And within the red square are the many ruined structures inside the great walls.

A medieval like drawbridge in the middle of a lunar crater? Turrets of various shapes? A lone standing column in full view? These are the objects that are clearly shown in the next photo.

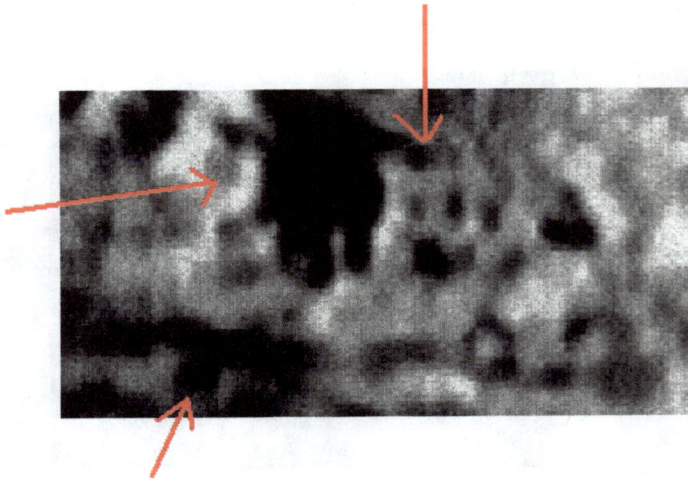

Inside the crater toward the front: The top arrow points down toward a drawbridge like structure with a turret on the left. The middle arrow points toward another castle like structure with an elbow like shaped turret pointing upward. And the bottom arrow points toward an opening, or gate, in a wall at the base of the crater. The earth like counterparts are shown below:

Front of two castles (earth and moon)

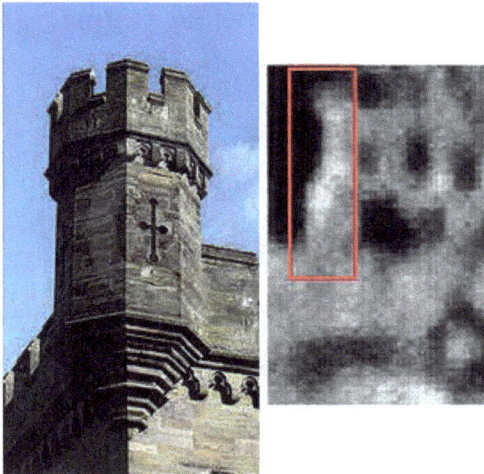

Turrets

Below is a wide view of the lunar drawbridge and the entire front area.

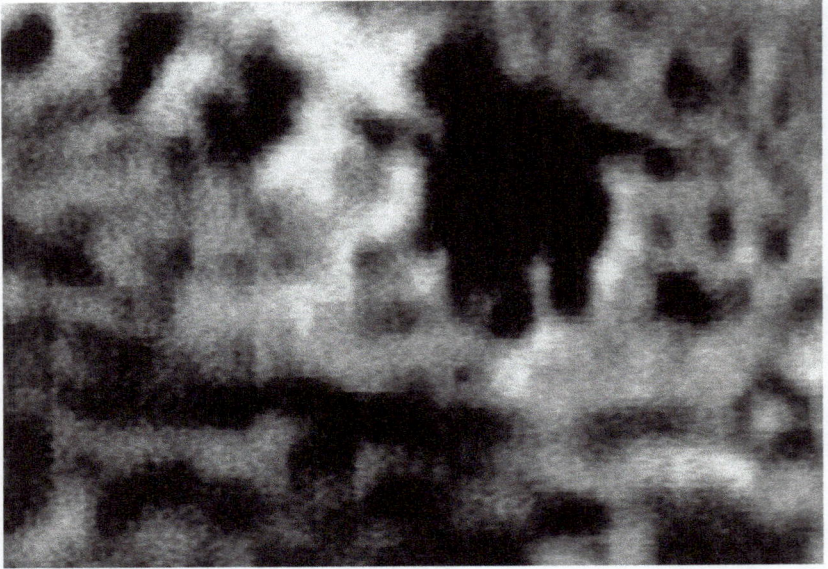

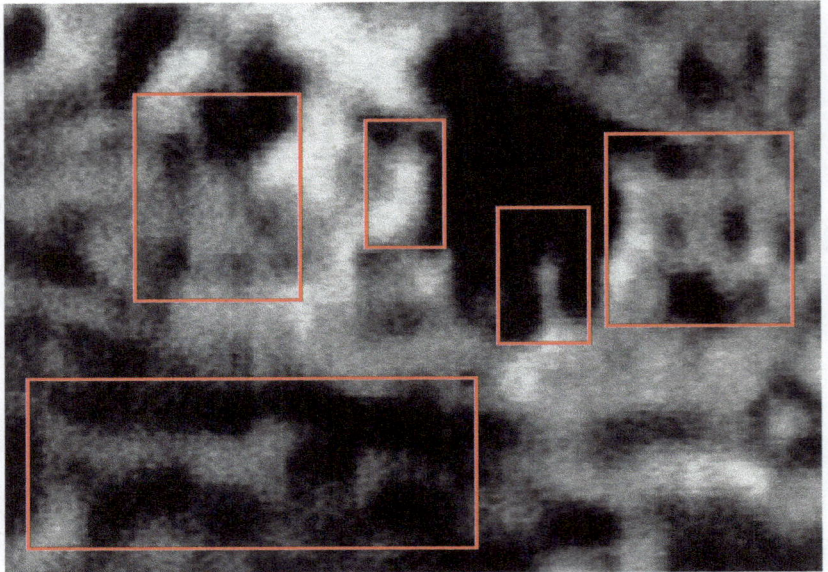

The areas in red are not what a person would expect in common crater material. The object on the extreme top left is a conical tower and it is right next to the elbow shaped turret. Of particular note is the simple upright column in the middle of the picture. Or is it a piece of statuary? How can that be explained away as simple crater material?

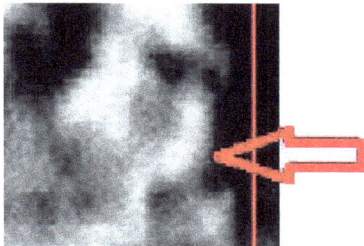

Stove pipe shaped turret?

Sculpted statuary?

This is not the simple inside of a naturally formed impact crater. It is the remains of a castle like structure with surrounding walls and an ancient city within. A person can almost envision a moat surrounding it. By the way, what is that dark area running along the base of the "castle?"

Between the city on the left and the castle within the crater is an even more unusual sight. Quite a remarkable object to be on the Moon. It has the appearance of a huge fountain similar to the one shown below, minus the water flow.

Next is what the Crater Eratosthenes Fountain looks like. Admittedly it is very difficult to define, however its general structure is clear. It is a curved object similar to a fountain. These types of shapes do not appear naturally on the Moon. There seem to be white boulders along either side of the fountain.

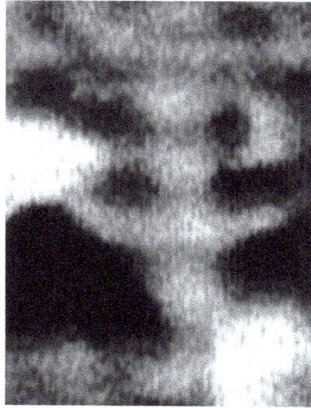

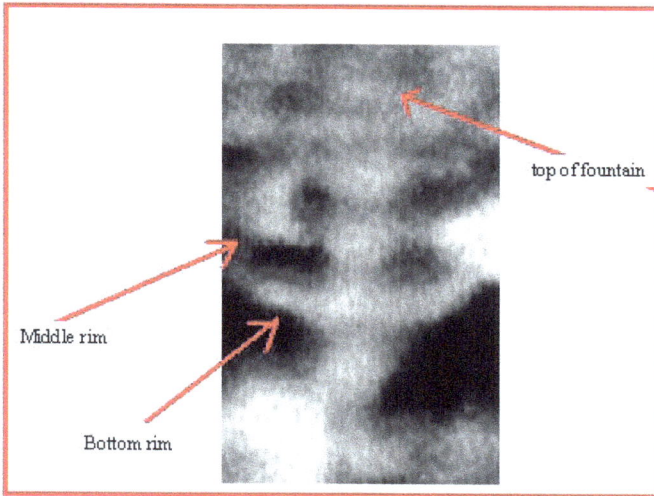

top of fountain

Middle rim

Bottom rim

It seems an item of small size to be able to photograph on the Moon but two things come into play. First is the probable actual huge size of this fountain; possibly more than a single story high. And secondly, the source photograph is one of the clearest and best ever taken from low orbit of this area under full sunlight.

The next picture is of particular interest because it lay directly between the grand fountain and the first true building of the ancient city proper. What is your first impression of this photograph?

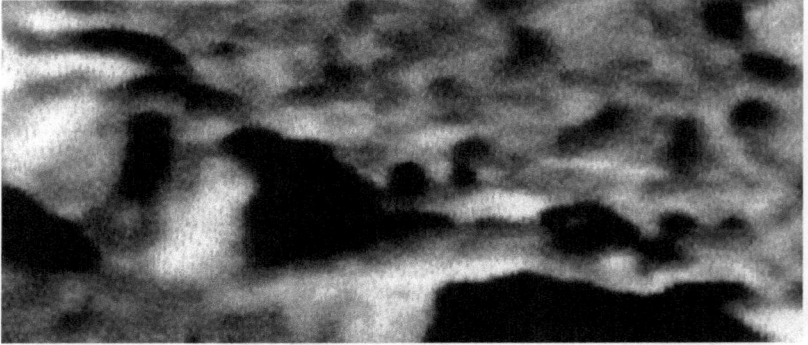

Does this not look like a graveyard! In the forefront – bottom right – is what looks to be a headstone lying on its back. In the background are small, low tombs. And on the far left is a large tomb with a gable over the front entrance. Below is a close up of this tomb.

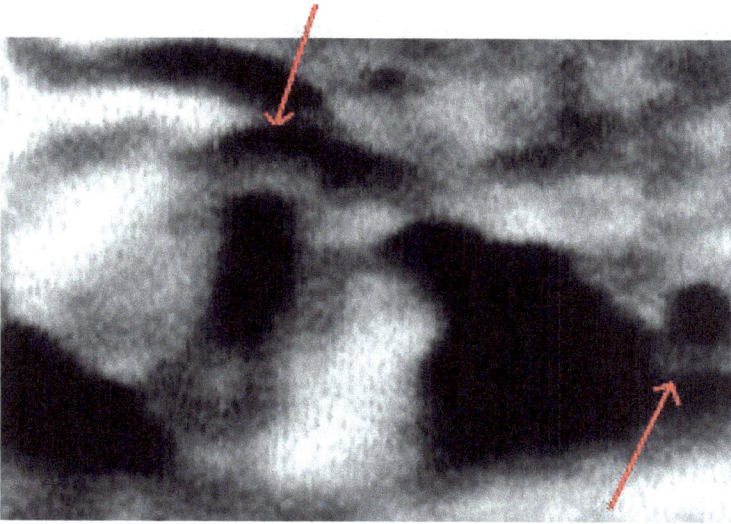

The arrow coming from the top points at the gabled roof over the doorway. And the arrow coming up from below? Does it not appear to be pointing at something like a tombstone?

Is there any wonder that through the ages and even up to the present time movement has been reported in the Erastosthenes area of the Moon? There is much here to evaluate and excavate by extraterrestrial archaeologists.

Before leaving the matter of Eratosthenes one more item needs to be seen. And the rather critical importance is why it is being presented last. Enough supporting evidence as to the validity of this site has been now given to make what is about to be put forth credible. Note what is within the red circle in the picture below.

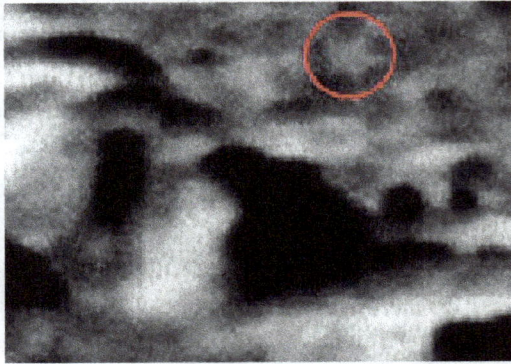

That is clearly the symbol of a star. And it is exactly the type of star that appears on the exteriors of planes and jets that are operated by the United States military in its various branches. It is highly significant that it appears in Selenite graveyard. Could this be a place set aside for the burial of the many persons of the American military who were lost while in the midst of pursuing or otherwise encountering UFOs? Beyond belief? There are claims of aliens being buried on earth in various locations, particularly in the town of Aurora, Texas. Why wouldn't the Selenites or their survivors reciprocate for downed humans?

From Eratsothenes we go to the Crater Aristarchus where the ancients built their city into the side of the crater wall in much the way the Pueblo Indians of the American Desert Southwest did.

Like Eratosthenes, Aristarchus has long been the sight of mysterious lights. In the 2005 photograph below, the entire crater can be seen aglow.

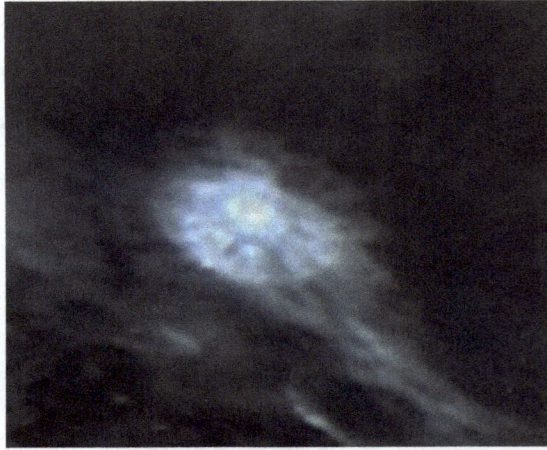

That's pretty impressive. The reason this study of Aristarchus is placed in this section rather than the one on alien activity is because I believe that the alien activity taking place here is more of an archaeological type. They are excavating the ancient site that fills one rim of the crater, the one part of the crater that is almost always deeply shaded in all of the photographs of it.

Aristarchus is another case of photographic tampering by NASA. NASA **wants** us to believe that this crater has been mined and shaved out by aliens. That is why they allow us to see the following extreme close up of the crater. It is a form of disinformation.

Or the other type of photograph they don't mind us seeing is the one where most of the crater is bleached by sunlight as below.

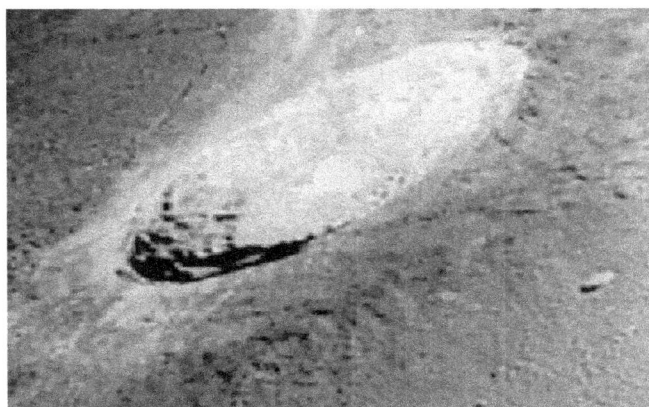

But even in this photo, part of the ancient city can still be seen as noted in the red square.

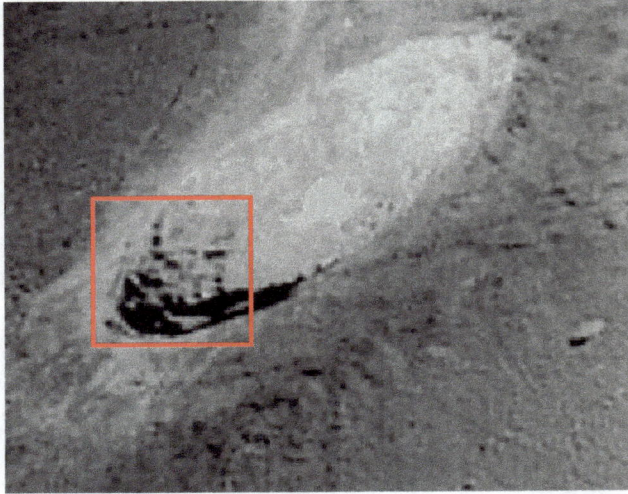

Looking directly downward on that section in red below is an enlarged version of it. It is a rather impressive structured layout of buildings whose roofs had long since vanished.

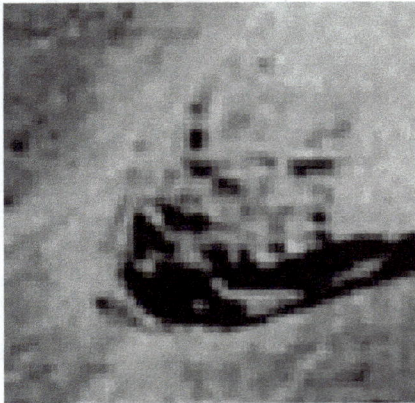

The areas in black are artificial structures not pixel deformations. And this is just one edge of the upper rim.

How was I able to get this picture in particular? As in most cases where anomalies have been discovered the images were made from older photographs taken at a time before NASA knew of the alien and Selenite presence on the Moon. Unfortunately for NASA, it was impossible for them to censor from the entire world the pictures of Aristarchus.

NASA simply made a mistake. Something which doesn't seem too uncommon for this organization. It allowed one of the Apollo photographs from orbit of Artistarchus to be released. It was actually part of a composite made up of several high resolution photographs and that is probably what contributed to NASA's confusion in that it was unaware how damaging one of these portions of the wider image would be. This is it below.

The enlarged view follows.

The above photograph is so damaging because it does not look anything like this:

But the next photograph is even a bigger problem for NASA. It reveals the cave like "rooms" built into the sides of the crater walls exactly like the Pueblo of the American Southwest.

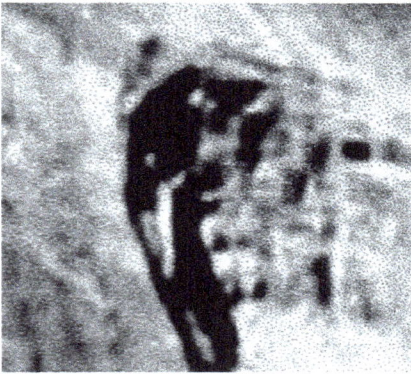

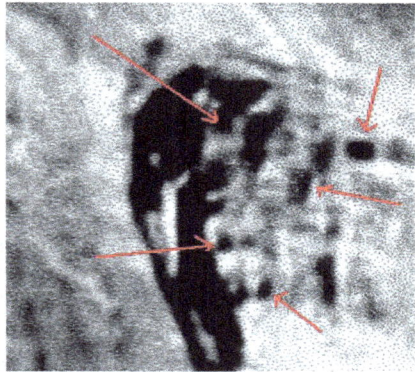

Pueblo cliff dwellings

Moving from Aristarchus, I refer to our next location as prime real estate. It is a small resort situated on the edge of the Grimadli Crater and to me it seems like a place where a wealthy Selenite of old might have gone on vacation or have owned a villa. It is located at the edge of the planet and gives a beautiful view of outer space beyond.

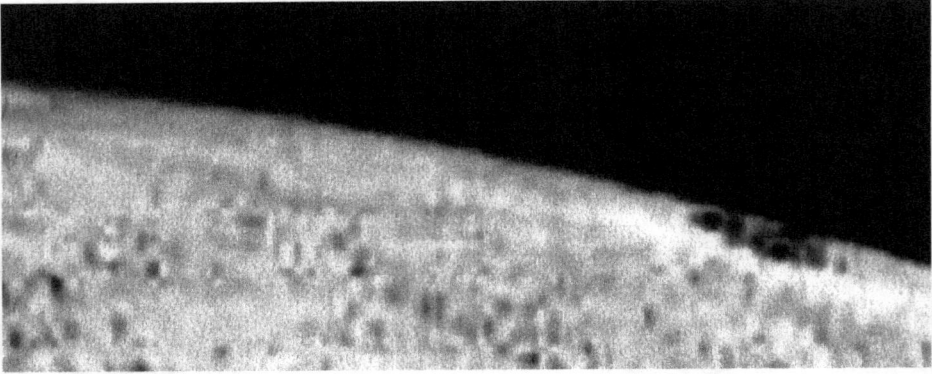

A closer view, then an enlargement of the "vacation homes" or rich Selenite's villas.

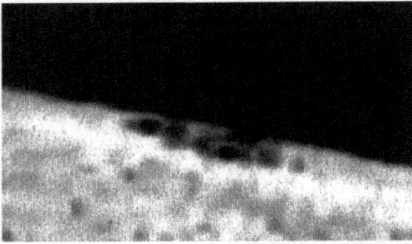

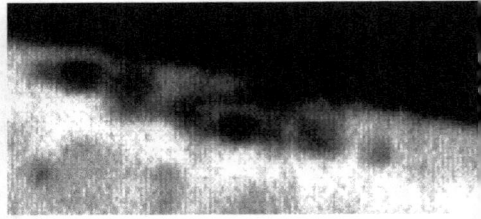

It isn't yet known who the inhabitants were – be they colonists from other worlds, or even from the ancient Earth – but that there once were thriving communities on this still busy world seems clear. Perhaps someday in the future the race who really controls the Moon will allow human beings back onto this world and archaeologists from earth will be able to perform their own excavations of these mysterious locations.

Lunar Apocalypse and Conclusions

It seems highly likely that one or more species of extraterrestrial is deeply invested in the Moon. A pair of 20th century Russian scientists have even published a paper in which they claim that the Moon is a gigantic spaceship that had been parked in earth orbit by aliens millennium ago.

More likely, however, is the hypothesis that the debris tossed into earth orbit by an asteroid strike later coalesced and became the Moon, developing into a habitable world long before its parent planet. Native life developed here first. Then the Moon may have been colonized by one or more alien races. When the aliens returned to the Moon to find that their colonies had died out due to drastic environmental change they decided to use the satellite as a base of operations from which to make frequent visits to a then much more hospitable earth. They may even have played a role in helping the development of the human species.

At any rate, based on the photographic evidence that has just been produced it is clear that the aliens continue to rule this part of our solar system and may be in strained communication with earth governments. The aliens most likely have bases on Mars, Titan and several other locations throughout the solar system.

Their relation with the earth civilization is like that of overlords to their inferiors. Remember the terror with which our astronauts viewed the aliens when they encountered them on the Moon. This type of relationship should be expected since the aliens are millions of years farther advanced than we.

It is important to keep in mind that there probably is more than one alien species operating on the Moon and not all of them are hostile. Among the aliens there appears to be at least one species who is concerned with the welfare of the human race even though the others may be somewhat unfriendly toward us. This benevolent race may in fact be genetically related to us.

Yet, we have been warned to discontinue our visits to the Moon. The reasons for this are not yet clear. But our government took the threats so seriously that it abandoned its planned final missions to the Moon in the 1970's even though they had already been funded and the crews had already been chosen. We should never forget that the Moon is ruled by a superior species of extraterrestrials and that we truly have no place we can hide or to which to escape should they decide to take over our world, too.

And this brings up the matter of a very grave threat to our world. As noted, much of the alien activity on the Moon has been underground. It's very likely that they have created a vast network of interweaving tunnels which may have caused great instability in the interior of the Moon. Apollo astronauts noted in their coded messages that they experienced the shock felt by underground blasts created by an unknown source, probably alien of origin.

It is very possible that at some point in the future the extraterrestrials will choose to takeover our planet and rid it of its entire population. This could be done by producing a fearsome detonation in the interior of the Moon and blowing the satellite into pieces. Most of these would hurtle toward the earth as gigantic asteroids and wreak havoc upon the planet. At this point the aliens could move their operations to earth and terraform the planet into whatever environment they choose. This would be the true Apocalypse that humankind has so long feared. Especially relevant is the Biblical description about the destruction of the Moon and mountains dropping into our seas. The only hope that humanity then would have would be Divine intervention.

The End

XXOC

www.ingramcontent.com/pod-product-compliance
Lightning Source LLC
Chambersburg PA
CBHW062144020426
42334CB00020B/2506